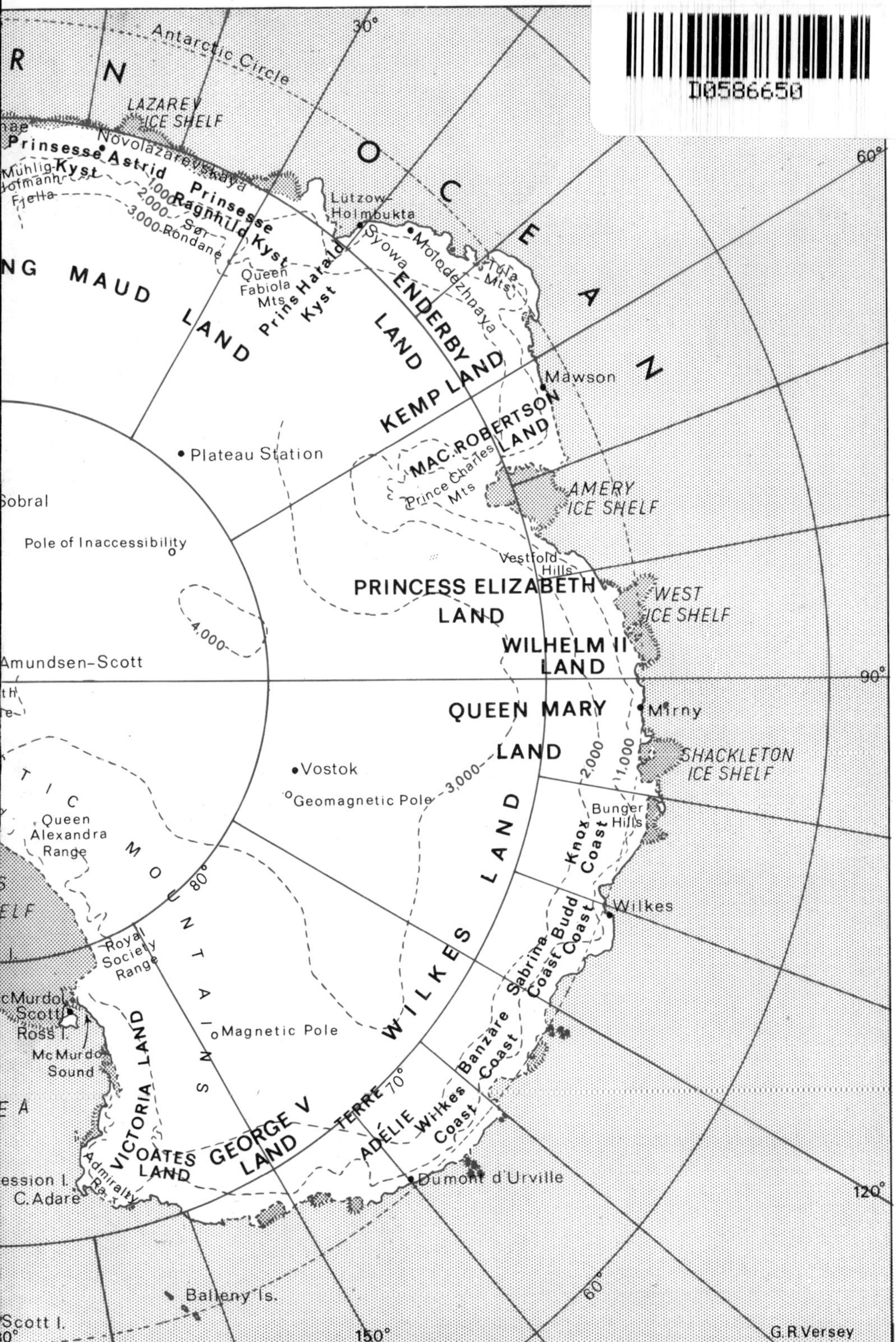
Antarctic Circle
LAZAREV ICE SHELF
Prinsesse Astrid Kyst
Novolazarevskaya
Muhlig-Hofmann Fjella
Prinsesse Ragnhild Kyst
1,000
2,000
3,000
Sør Rondane
Queen Fabiola Mts
Prins Harald Kyst
Lutzow-Holmbukta
Syowa
Molodezhnaya
Tula Mts
O C E A N
MAUD LAND
ENDERBY LAND
KEMP LAND
Mawson
MAC. ROBERTSON LAND
Prince Charles Mts
AMERY ICE SHELF
Plateau Station
Sobral
Pole of Inaccessibility
Vestfold Hills
PRINCESS ELIZABETH LAND
WEST ICE SHELF
4,000
WILHELM II LAND
Amundsen-Scott
QUEEN MARY LAND
Mirny
SHACKLETON ICE SHELF
Vostok
Geomagnetic Pole
Bunger Hills
Knox Coast
Queen Alexandra Range
Wilkes
Sabrina Coast
Budd Coast
WILKES LAND
Royal Society Range
McMurdo
Scott
Ross I.
McMurdo Sound
Magnetic Pole
Banzare Coast
VICTORIA LAND
OATES LAND
GEORGE V LAND
TERRE ADÉLIE
Wilkes Coast
Dumont d'Urville
Admiralty Ra.
C. Adare
Balleny Is.
Scott I.
30°
60°
90°
120°
150°
80°
70°
60°
G.R.Versey

THE ANTARCTIC

For St. Paul's School
Library with
H.G.R. King's
best wishes
3.12.76

THE ANTARCTIC

by H. G. R. KING

BLANDFORD PRESS
LONDON

First published in 1969

167 High Holborn, London W.C.1

SBN 7137 0373 3

Text filmset in 11 on 12 point Baskerville
and printed and bound in Great Britain by
Jarrold and Sons Ltd, Norwich

CONTENTS

LIST OF COLOUR ILLUSTRATIONS

AUTHOR'S PREFACE

In December 1959 twelve nations acceded to the Antarctic Treaty '. . . Recognizing that it is in the interest of all mankind that Antarctica shall continue to be used exclusively for peaceful purposes and shall not become the scene or object of international discord'. In the years that have followed this historic declaration we have seen a rising level of scientific research in the treaty area with studies spreading into many fields. These complex international research programmes will continue for the foreseeable future to add to the basic knowledge of man's environment.

One effect of this increasing activity has been an explosion in the scientific and technical literature whose shock waves are rocking polar libraries. Of this vast output of Antarctic publications only a fraction seems to be known or accessible to the man-in-the-street whose interest and enthusiasm in the nature and history of the Antarctic regions seems to be equally in the ascendant. Evidence for this is to be found in the files of the Scott Polar Research Institute, where, as Librarian and Information Officer, I handle annually hundreds of enquiries from people all over the world seeking information on the Antarctic. Their need, it seems to me, is for single comprehensive reference books, plain men's guides written in non-technical language, amply illustrated and covering, in outline at least, the main facets of the Antarctic scene.

The 1960s have seen the publication of two major English language reviews of Antarctic research – *Antarctic research* (1964) and *Antarctica* (1965). Both are collections of contributions by leading specialists and are indispensable to the serious student. There have, too, been a number of less specialized books valuable as expositions of Antarctic science at popular level but omitting much of the basic geographical and historical background material for which I am so frequently asked. I have compiled *The Antarctic* to meet what I feel to be a very real need. It aims to bring together much scattered material relating to the Antarctic continent, the surrounding Southern Ocean, the Antarctic islands and the adjoining sub-Antarctic, the last a region notably neglected in the general literature but of increasing interest to science. My hope is that this book will not only serve as an introduction to Antarctic science but that it will provide the general reader with some of the basic facts concerning Antarctic geography, natural history and exploration.

With regard to the highly specialized subjects which I have tried to interpret I must here acknowledge my very considerable debt to the two books referred to above, as well as to numerous other books and articles. In addition I would like to acknowledge the aid of the specialists who have read specific chapters thereby saving me from spectacular disaster in one technical crevasse after another. Naturally I take full responsibility for all errors and omissions. My especial thanks are due to Dr. T. E. Armstrong, Dr. S. Evans, Dr. P. F. Friend, Dr. S. W. Greene, Dr. B. B. Roberts, Dr. G. de Q. Robin, Mrs. Ann Shirely (neé Savours) and Dr. C. W. M. Swithinbank. I wish additionally to thank Mr. G. R. Versey for drawing maps, and my wife and Mrs. A. Proudman for reading proofs, Mrs. J. Blake and Miss K. Hollick for typing drafts, and my friends from the staff of the Blandford Press, Miss V. Frampton, Miss R. Ketley and Mr. R. R. Knightley for guiding me down the labyrinthine corridors of book production.

H. G. R. K.

ACKNOWLEDGEMENTS

Acknowledgement is due to the following individuals and organizations whose illustrations are reproduced in the book:

R. J. Adie and *New Scientist*, pp. 60, 61
R. J. Adie and *Science Journal*, p. 68 (*2*)
A. C. Bibby, title page
G. Blundell, Pls. 10, 11
D. Borthwick, p. 157 (*2*)
Robert Clarke, p. 128 (*bottom: photo and copyright*)
J. T. Darby, p. 54 (*bottom*)
L. Davies, p. 10 (*top*)
R. Délepine, Pl. 45
S. W. Greene, p. 152, Pls. 15, 17, 19, 20, 21, 22
J. Linsley Gressitt, pp. 105 (*centre and bottom*), 106 (*2*)
M. W. Holdgate, Pls. 38, 39, 41, 42
George Holton, pp. 20, 21 (*3*)
Frank Hurley, 222 (*top right*)
R. E. Leech, p. 105 (*top*)
R. E. Longton, Pls. 16, 18
P. Moiseyev, pp. 128 (*top*), 134 (*bottom*)
David Petrie, pp. 46 (*top right and bottom left*), 191, Pls. 30, 51, 54, 59
E. Picciotto, p. 13 (*bottom*)
Brian Roberts, pp. 11 (*top*), 117 (*bottom*), 120 (*right*), 121 (*centre and bottom*), 140 (*bottom left and bottom right*), 162 (*top*), 179 (*top*), Pls. 24, 27, 28, 43, 44, 46, 47
Charles Swithinbank, pp. 10 (*bottom*), 26, 57, 95, 110 (*top*), 116 (*centre*), 125 (*bottom*), 179 (*bottom*), 183 (*bottom*), 185 (*bottom*), 187 (*bottom*), 190, 199 (*top and centre*), Pls. 2, 3, 4, 6, 7, 13, 23, 31, 32, 33, 36, 48, 50, 52, 53, 55, 56, 58, 60, 61, 62, 63, 64
G. R. Versey, *endpapers*, pp. 3, 4, 34, 72, 82
Y. Yoshida, p. 178 (*centre*)
American Geographical Society, p. 45 (*bottom 2*)
Arctic Institute of North America, Washington, D.C., p. 195

Associated Press Ltd., p. 247 (*right*)
Australian Department of Supply, Antarctic Division, pp. 54 (*top right*), 78 (*bottom left*), 163, 166 (*top and bottom*), 178 (*bottom*), 199 (*bottom*)
British Antarctic Survey, pp. 16, 25 (*2*), 42 (*top*), 43, 46 (*top left*), 59, 63, 78 (*top left, top centre, top right, bottom right*), 90, 91, 101 (*top left*), 110 (*bottom*), 117 (*top*), 120 (*top left*), 141 (*bottom*), 155, 156, 183 (*top*), 184 (*top left, top right, centre left, centre right*), Pls. 1, 5, 9, 12, 14, 25, 26, 34, 35, 40, 49, 57
British Graham Land Expedition, p. 226
Camera Press Ltd., p. 118 (*bottom*)
Dominion Museum, New Zealand, p. 116 (*top*)
Expéditions Polaires Françaises, p. 187 (*top*)
Gough Island Scientific Survey, Pl. 29
H.M. Stationery Office, Crown Copyright, reproduced by permission, pp. 29 (*2*), 30, 127, 153 (*2*)
Hunting Surveys Ltd., p. 55 (*centre*)
Illustrated Newspapers Ltd., p. 177
Japanese Antarctic Research Expeditions, p. 55 (*top*)
McGraw-Hill Book Company, p. 32
National Institute of Oceanography, pp. 28, 125 (*top*)
National Science Foundation, Washington, D.C., pp. 76, 101 (*top right*), 107, 108, 110 (*centre*), 116 (*bottom*), 118 (*top*)
National Science Museum, Department of Polar Research, Tokyo, p. 169
New Zealand Department of Scientific and Industrial Research, pp. 24, 138 (*left*), 166 (*centre*), 175 (*2*), 179 (*centre*), 194, 218 (*2*), 222 (*bottom*)
Norwegian-British-Swedish Antarctic Expedition, pp. 51, 232, 233
Novosti Press Agency, p. 42 (*bottom*)
Scott Polar Research Institute, pp. 10 (*centre: S. A. Shingler*), 45 (*top*), 84, 98, 134 (*top and centre*), 141 (*top*), 143 (*bottom left and bottom right*), 171, 179 (*top*), 184 (*bottom*), 198, 203, 206, 210, 216, 219 (*3*), 220 (*2*), 222 (*top left*), 230 (*right*), 238–9, 244 (*2*), 276, Pls. 65A, B, C
South African Weather Bureau, p. 162 (*bottom*)
Terres Australes et Antarctiques Françaises, Paris, p. 138 (*right*)
Trans-Antarctic Expedition, pp. 58, 245 (top), 247 (left)
United States Antarctic Expedition, p. 228 (*3*)
United States Geological Survey, pp. 6, 36 (*top: TMA 549, Fan 31, Exposure 206*), 37 (*top: TMA 541, Fan 33, Exposure 69*), 40 (*TMA 923, Fan 33, Exposure 235*), 52 (*top: TMA 570, Fan 33, Exposure 264; bottom: TMA 566, Fan 33, Exposure 176*), 53 (*TMA 550, Fan 33, Exposure 217*), 65 (*2*)
United States Naval Support Force, Antarctica, *facing* p. 1
United States Navy, pp. 11 (*bottom*), 13 (*top*), 27, 35, 36 (*centre and bottom*), 37 (*bottom*), 46 (*bottom right*), 54 (*top left*), 55 (*bottom*), 101 (*bottom*), 123, 172 (*2*), 173, 174, 185 (*top and centre*), 188 (*3*), 189 (*3*), 231 (*2*), 243, 245 (*bottom*), 251
The illustration on p. 193 is reprinted by permission of G. P. Putnam's Sons from *90° South* by Paul Siple. Copyright © 1959 by Paul Siple

Acknowledgement is also due to the following:

T. W. Bagshawe and the Cambridge University Press for permission to quote from *Two Men in the Antarctic*
William Heinemann for permission to quote from Douglas Mawson's *The Home of the Blizzard*
Dr. B. B. Roberts and the Editor of *Oryx* for permission to quote from the former's article in *idem*, Vol. 8, No. 4, 1966, pp. 237–43

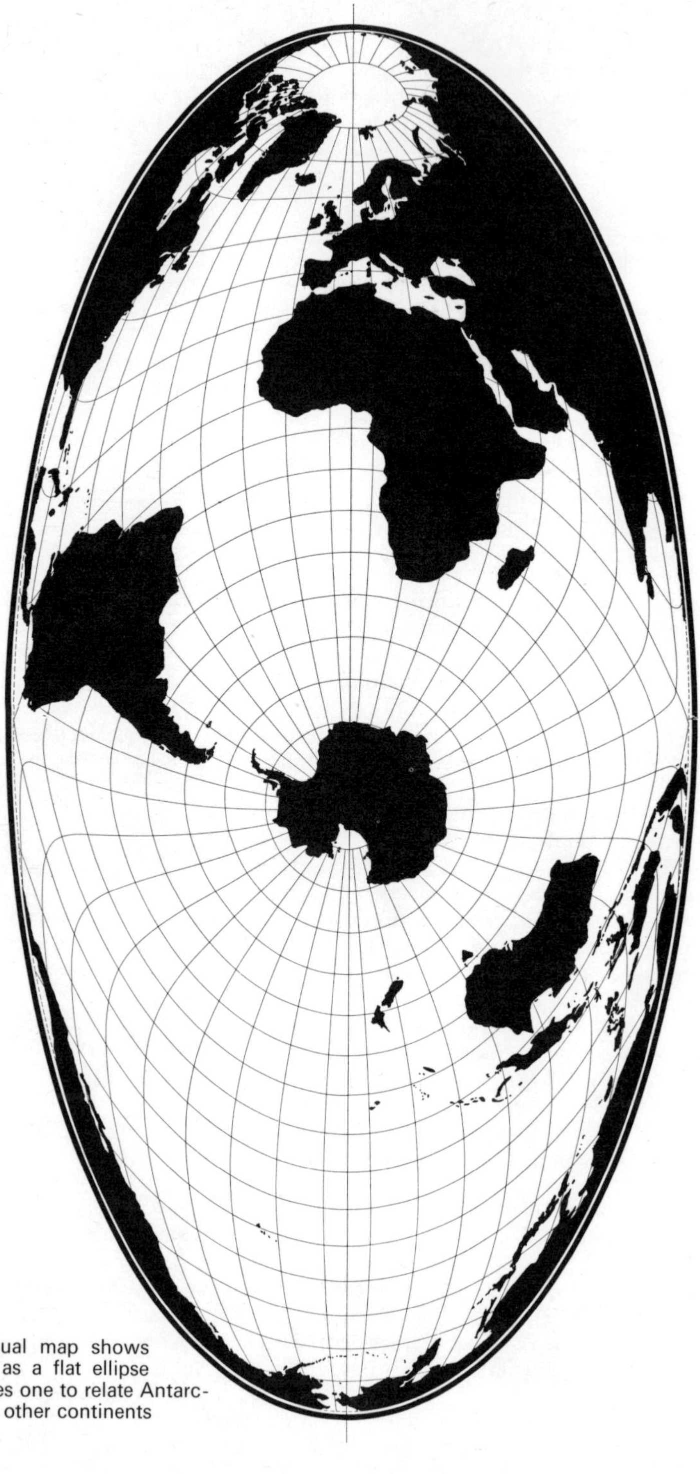

This unusual map shows the Earth as a flat ellipse and enables one to relate Antarctica to the other continents

I

INTRODUCING THE ANTARCTIC

It was the ancient Greeks who gave us the word for it. The polar regions, long before they came to be explored by man, were named with terms derived from the Greek cosmography. Our word Arctic comes from *arktos*, meaning The Bear, the name given by the Greeks to the constellation which rotated about the North Polar regions. Eventually the term came to be applied to the sea and land regions surrounding the North Pole itself and its derivative, Antarctic (literally 'opposite the Arctic'), to the regions round the South Pole. The Antarctic regions, usually referred to simply as the Antarctic, do not lend themselves to exact specification, indeed the term is left deliberately vague. Specialists in various scientific disciplines each tend to use definitions which are valid enough for their own special purposes. This opening chapter, therefore, is not concerned with drawing any firm Antarctic boundary round the map of the Southern Hemisphere. Rather we shall look at a few existing definitions as a guide to setting our own limits to the Antarctic for the purposes of this book. We can then proceed to outline some of the region's natural features and point to a few of man's activities in this isolated, hostile but challenging part of the world.

Antarctic boundaries

The focal point of the Antarctic regions is the large and ice-covered continent called Antarctica centred asymmetrically on the South Geographical Pole. Surrounding Antarctica is a vast ocean, the Southern Ocean, also partly ice-covered, whose northern boundary approximates to a line joining the tips of the surrounding continents. The emptiness of the Southern Ocean is relieved by a few widely scattered islands and island groups insignificant in area but valuable as observatories in this waste of ice and water. Northwards of the pack ice towards the trade wind belts – the 'roaring forties' and the 'filthy fifties' – lie more islands with less rigorous climates, a region known as the sub-Antarctic. Establishing a boundary to any natural region is inevitably arbitrary as changes in climate, life and scenery tend to be gradual without sharp limits. We can illustrate the problem by considering first of all the polar circles, familiar to many as the two lines of latitude drawn round the globe at 66°33′N. and S.; this is the angle which the Earth's axis makes with the plane of the Earth's orbit. We are accustomed to thinking of the polar circles as, by definition, enclosing the Arctic and Antarctic regions. They are very frequently defined as those areas of the Earth's surface where the Sun does not sink below the horizon at least once during the year and also does not rise above the horizon on at least one day a year. Theoretically this might appear to be true, but in practice, because of refraction of its rays, the Sun is still visible for a few minutes on every day of the year as far as 60 miles within the polar circles and stays visible on the horizon, at midnight, on at least one day a year to a distance of 60 miles outside them. Again, theory has it that as one approaches the poles the length of the period of constant sunshine, or darkness, increases so that at the poles themselves night and day last six months each. But the practical experience of travellers contradicts this; the effect of refraction is enough to give the poles themselves approximately half a week more than six months with sunshine and half a week less without. So much for some misconceptions about the polar circles. As natural boundaries they are quite useless; a glance at the map will show that much of the peninsula region of the Antarctic continent lies to the north of the Antarctic Circle.

A more obvious feature which suggests itself as a boundary to the Antarctic – and it is a very real obstacle to ships – is the edge of the floating pack ice. But the position of this is never certain since it alters in position from season to season and from year to year; as a basis for delimiting the Antarctic it too is unsatisfactory.

A boundary that has gained general acceptance, and to which we shall refer constantly throughout this book, is an oceanic one. It cannot be seen, of course, but the southward-bound traveller can feel it as a measurable drop in temperature. Oceanographers term it the Antarctic Convergence, a belt of water some 20 to 30 miles wide where the cold northward-flowing currents stemming from the Antarctic continent sink beneath warmer southward-flowing water. This belt of mixing water girdles the entire Southern Ocean, varying in average position but roughly following the 50th parallel of latitude through most of the Atlantic and Indian Ocean sectors and located between latitudes 55° and 62° in the Pacific sector. The Antarctic Convergence marks not only a change in the ocean's surface temperature but a change in chemical composition as well. There are also marked biological differences on either side of the convergence not only in the plants and creatures inhabiting the ocean but among the bird life also. Finally, the convergence influences the climate of the islands whose shores are washed by its cold upwelling surface waters, determining their characteristics which are predominantly those of the Antarctic continent.

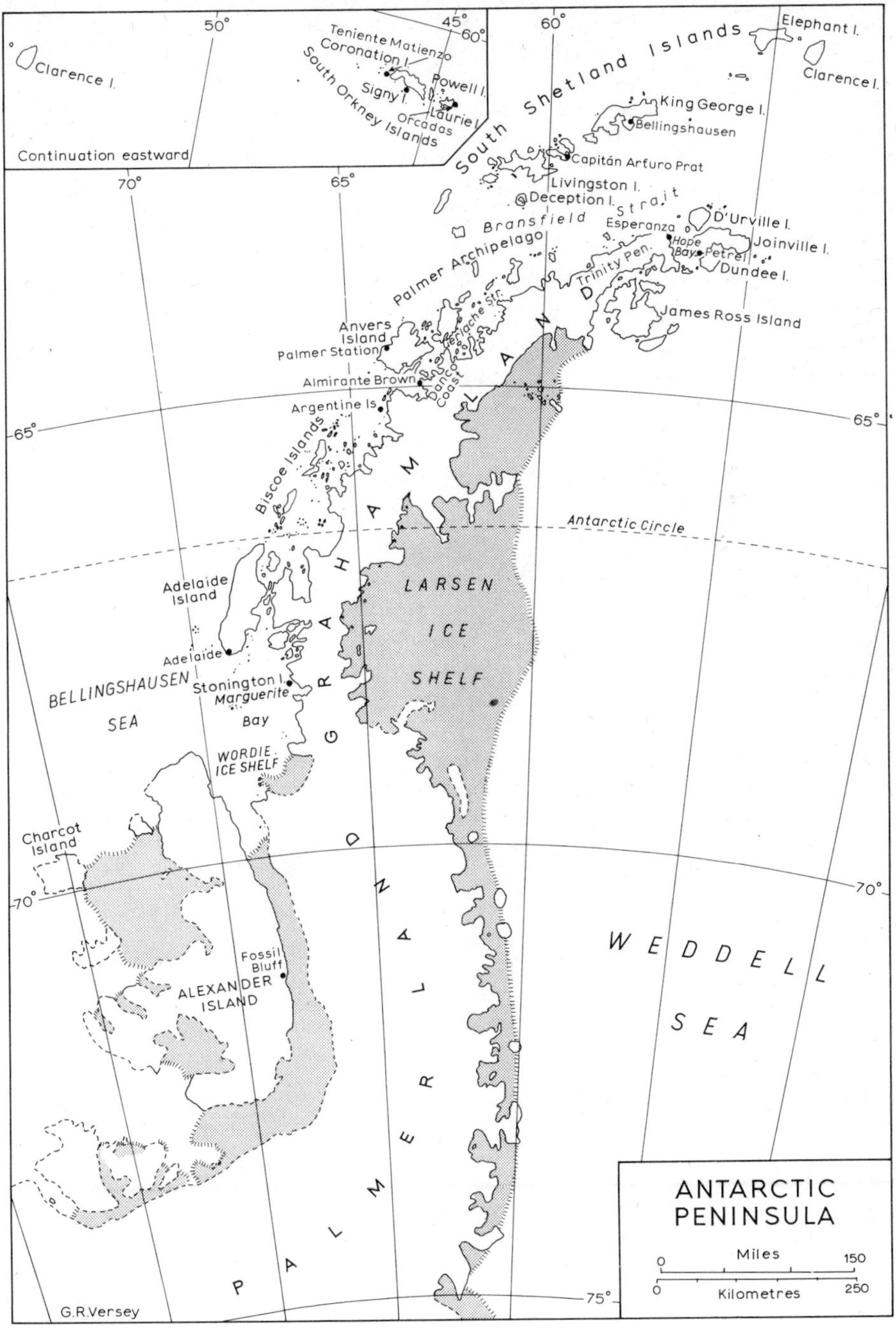

Continuation eastward
Clarence I.
South Orkney Islands
Teniente Matienzo
Coronation I.
Signy I.
Powell I.
Laurie I.
Orcadas
South Shetland Islands
Elephant I.
Clarence I.
King George I.
Bellingshausen
Capitán Arturo Prat
Livingston I.
Deception I.
Bransfield Strait
Palmer Archipelago
Esperanza
Hope Bay
Petrel
D'Urville I.
Joinville I.
Dundee I.
Trinity Pen.
James Ross Island
Anvers Island
Palmer Station
Gerlache Str.
Danco Coast
Almirante Brown
Argentine Is.
Biscoe Islands
GRAHAM LAND
Antarctic Circle
Adelaide Island
Adelaide
LARSEN ICE SHELF
BELLINGSHAUSEN SEA
Stonington I.
Marguerite Bay
WORDIE ICE SHELF
Charcot Island
Fossil Bluff
ALEXANDER ISLAND
PALMER LAND
WEDDELL SEA
ANTARCTIC PENINSULA
Miles
0
150
0
Kilometres
250
G.R.Versey

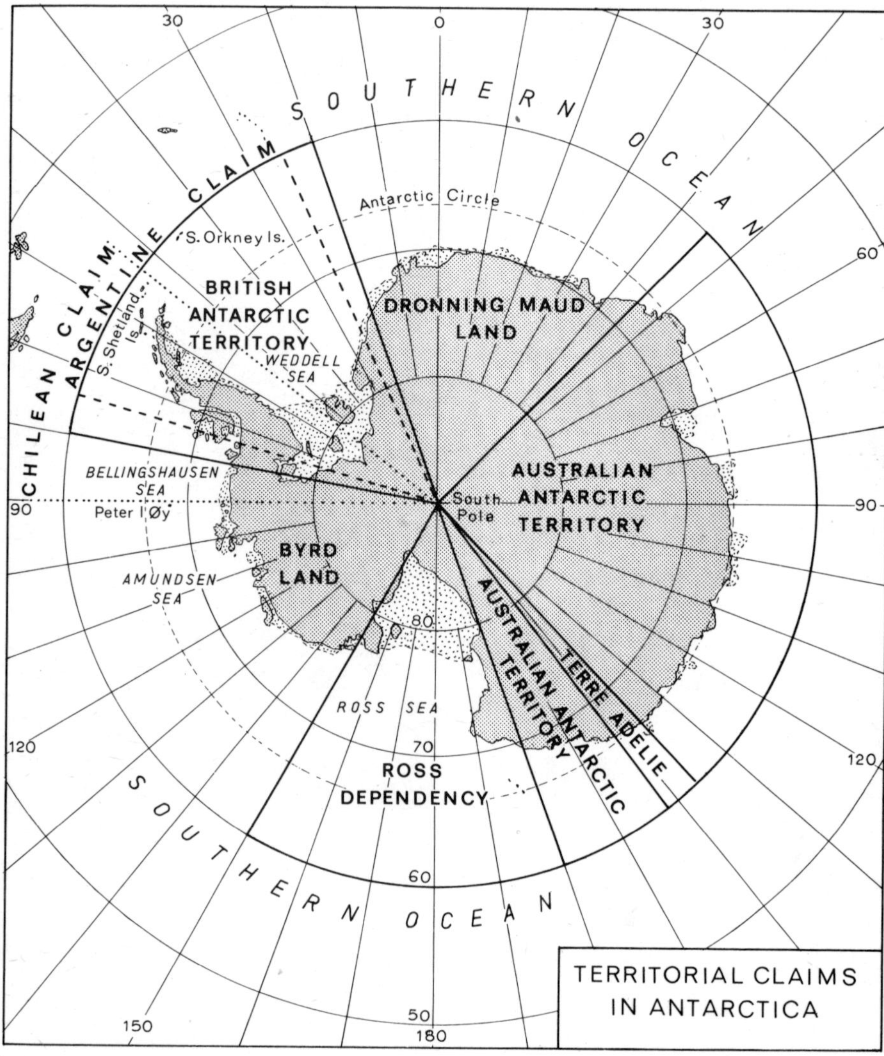
SOUTHERN OCEAN
ARGENTINE CLAIM
CHILEAN CLAIM
S. Orkney Is.
S. Shetland Is.
BRITISH ANTARCTIC TERRITORY
WEDDELL SEA
Antarctic Circle
DRONNING MAUD LAND
AUSTRALIAN ANTARCTIC TERRITORY
South Pole
BELLINGSHAUSEN SEA
Peter I Øy
BYRD LAND
AMUNDSEN SEA
ROSS SEA
ROSS DEPENDENCY
AUSTRALIAN ANTARCTIC TERRITORY
TERRE ADÉLIE
SOUTHERN OCEAN
30
0
30
60
90
90
120
120
150
180
80
70
60
50
TERRITORIAL CLAIMS IN ANTARCTICA

Throughout this book we shall refer to the region south of the Antarctic Convergence as Antarctic and that to the north of it as sub-Antarctic. The northern boundary of the sub-Antarctic region is also an oceanographic one, the sub-tropical Convergence, which, as the map shows, verges on the sub-tropics. The sub-Antarctic is a region of increasing interest to biologists, its plant and animal life having many Antarctic affinities.

For present purposes we can now define the Antarctic regions as all the seas and lands lying within the region to the south of the Antarctic Convergence, namely the continent of Antarctica, the southern part of the Southern Ocean and the following islands: the South Shetland Islands, the South Sandwich Islands, the South Orkney Islands, South Georgia, Bouvetøya, Heard Island, the Balleny Islands, Scott Island and Peter I Øy. We shall also count as Antarctic Archipel de Kerguelen, Iles Crozet and the Prince Edward Islands. These latter, though they lie north of the convergence, are near enough to its cold currents to have acquired Antarctic characteristics. Further north are a number of sub-Antarctic islands which have some features shared in common with the true Antarctic; they include the Falkland Islands, Tristan da Cunha; Gough Island, Ile St. Paul, Ile Amsterdam, Macquarie Island, McDonald Island and the sub-Antartic islands of New Zealand such as the Auckland Islands and Campbell Island.

With a bird's-eye picture of the region before us we can now take a closer look at the Antarctic scene; in nearly every respect we shall find that it is essentially different from any other part of the Earth's surface.

Antarctic characteristics

Antarctica, the continent which dominates the Antarctic regions, is 5½ million square miles in area. Roughly the equivalent of the U.S.A. and Mexico put together, it is by no means the largest of the continents. To most people it is as remote and mysterious as the Moon; indeed to many the Moon may well seem more familiar if no less accessible. For Antarctica is truly remote in terms of distance. The nearest mainland is the tip of South America, some 600 miles from the Antarctic Peninsula, but separated from it by Drake Passage, one of the world's stormiest stretches of ocean. More distant still are New Zealand (2100 miles) and South Africa (2500 miles). The major centres of population in the Northern Hemisphere are even further removed; the 8000 miles sea voyage from Southampton to the South Shetland Islands takes well over a month. Other continents are by contrast relatively neighbourly; only 56 miles separate North America from Asia at Bering Strait; Australia is only 40 miles from Eurasia at its nearest point. Add to sheer physical distance the fact that Antarctica is surrounded for much of the year by a barrier of floating pack ice, navigable to shipping in only a few select places during a short summer season, and it is understandable why there is no southerly equivalent of the Eskimo on

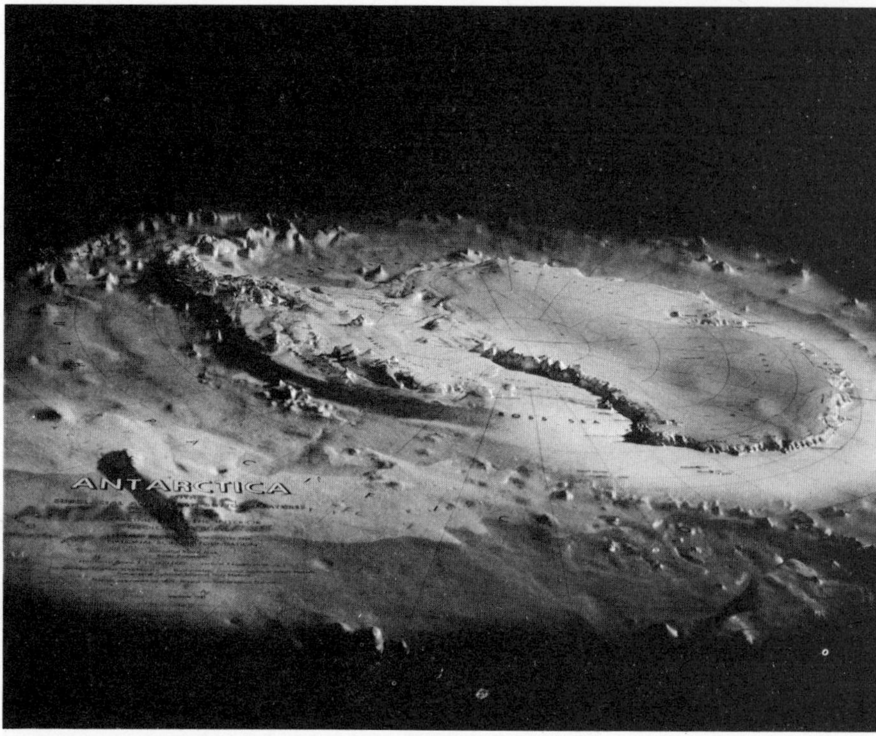

Photograph of a model of Antarctica showing features beneath the ice and ocean. The form of the bedrock beneath the ice has been found from numerous ice-thickness measurements

the continent and why the sum total of those privileged to have admired the splendours of the Antarctic landscape number only a few thousand.

Antarctica is remarkable for the nature of its outline; roughly circular in shape it is almost entirely contained within the circumference of the Antarctic Circle. This regularity is broken only by the lengthy Antarctic Peninsula, some 800 miles from base to tip, which points finger-like towards distant South America, and by the two great indentations of the Southern Ocean, the Ross and Weddell Seas. Two minor indentations have also been given the names of Amundsen Sea and Bellingshausen Sea respectively.

The major feature of Antarctica is the extensive ice-sheet which covers 95 per cent of the total area; mountains, valleys and plains are all hidden from view by ice averaging between 6500 ft. and 12,000 ft. in thickness. Bare rock will be found only on the exposed peaks (or *nunataks*) of the higher mountain ranges and at a few sheltered places round the coast. Locked up in Antarctica is over 90 per cent of all the snow and ice in the world. This ice-sheet is virtually one huge glacier of continental proportions, comparable only with the one that covers much of Greenland. As the ice has slowly piled up over many thousand of years it has, under its own sheer weight, flowed downwards and outwards towards the coasts. Here in places it has spread over the surface of the sea to form floating ice-shelves. Ice-shelves are a distinctive feature of Antarctica; two of the largest are the Ronne and Ross Ice-Shelves, the last covering a region larger than France. From time to time the seaward edges of these ice-shelves break off to form the flat-topped tabular icebergs which are so typical of Southern Ocean waters. Submerged under ice Antarctica serves to remind us of what northern Europe and North America must have looked like during the Pleistocene Ice Age a million years ago.

Though much of the original landscape of Antarctica is lost to view, modern technology has equipped the scientist with ingenious devices for detecting its subglacial hills and valleys. We know now that Antarctica is divided both geographically and geologically into two natural provinces; one of these, Lesser Antarctica, includes the peninsula and Byrd Land; the other, Greater Antarctica, embraces the bulk of the continent and is largely a region of high plateau. The dividing-line between the two is one of the world's greatest mountain chains, the Transantarctic Mountains. Alternative names for Lesser and Greater Antarctica are respectively West and East Antarctica, referring to the regions to the west and east of the 0°–180° Greenwich meridian which bisects the continent. These names can become rather meaningless so close to the Pole and can be the cause of some misunderstanding. Greater Antarctica is twice as large as Lesser Antarctica and geologically much older; it is what geologists call a *shield*, a term used to describe a very old and stable part of the Earth's crust, and is similar to the continental shields of Africa, Australia and Canada. In Greater Antarctica the shield forms a high plateau rising to over 13,000 ft. above sea-level. Elevation is indeed one of Antarctica's principal topographical features and it is on average the highest of all the world's continents. If the ice were to be removed from Greater Antarctica much of the underlying land would remain above sea-level. Lesser Antarctica is geologically very different; it is composed of a series of mountain ranges whose rocks are much younger. The mountains of the Antarctic Peninsula, scenically one of Antarctica's most attractive regions, are considered to be a continuation of the Andes of South America to which they are linked by a submarine arc that curves out to enclose the Scotia Sea; the South Orkney Islands, the South Sandwich Islands and South Georgia are but the surfacing peaks of this undersea chain. The mountains of Lesser Antarctica are less elevated than those of Greater Antarctica and if the ice here were to melt only a number of islands would remain. The land of Antarctica, therefore, is strictly a continent plus an archipelago. But this is only a theoretical situation. The all-embracing ice is very real and very permanent and we therefore may regard it simply as one very large continental area.

Ice is clearly linked with climate. In Antarctica we cannot be sure

THERMOMETER COMPARISONS

Centigrade	Fahrenheit	Centigrade	Fahrenheit
100°	212°	25°	77°
95	203	20	68
90	194	15	59
85	185	10	50
80	176	5	41
75	167	0	32
70	158	−5	23
65	149	−10	14
60	140	−15	5
55	131	−17·8	0
50	122	−20	−4
45	113	−25	−13
40	104	−30	−22
35	95	−35	−31
30	86	−40	−40

To reduce Fahrenheit to Centigrade, subtract 32 degrees and multiply by $\frac{5}{9}$; to reduce Centigrade to Fahrenheit, multiply by $\frac{9}{5}$ and add 32 degrees.

LINEAR COMPARISONS

1 millimetre = 0·03937 inch
1 centimetre = 0·3937 inch
1 decimetre = 3·937 inches
1 metre = 39·370113 inches = 3·280843 feet = 1·0936143 yards
1 kilometre = 0·62137 mile

1 inch = 2·54 centimetres
1 foot = 30·48 centimetres
1 yard = 0·914399 metre
1 mile = 1·6093 kilometres

CONVERSION TABLES
(For use throughout the book)

whether the climate perpetuates the ice or the ice perpetuates the climate. What is beyond argument is that some places in Antarctica are the coldest on Earth. On the high plateau of Greater Antarctica temperatures as low as −88.3 °C have been recorded, and the thermometer never rises above zero. At the South Pole the average temperature is about −50 °C. Over the continent as a whole there is a considerable degree of variation. Average temperatures for the west coast of the Antarctic Peninsula are around −3 °C with at least one month in the year when the average rises above freezing-point. But taken overall Antarctic temperatures are low – considerably lower than average Arctic temperatures. Precipitation is slight. At the South Pole, which is situated on a high plateau about 10,000 ft. above sea-level, the air is so cold and so dry that rain never falls and less than 2 in. of snow is measured annually. Incidentally, Antarctic snow is not the familiar fluffy variety but dry and granular like fine sand. Near the coast the snowfall is rather heavier and there may even be some rain in places. The actual feeling of being cold is very much a factor of the wind; Antarctic winds are notorious for their strength and persistence. When cold air from the high Antarctic ice-sheet comes cascading down the coastal slopes to sea-level and there coincides with cyclonic winds blowing in from the sea, the result is a hurricane with winds exceeding 100 m.p.h. The drifting snow and the blizzards which result from these constant strong winds are a major hazard facing the explorer of the Antarctic continent. Low average temperatures, low precipitation, high winds – these basic climatic features of Antarctica are to be found on the surrounding islands also, though conditions gradually ameliorate with distance from the continent. Temperatures on such islands as South Georgia, Heard Island, Archipel de Kerguelen and Macquarie Island are above the freezing-point for at least half the year round the coasts. Rain, too, is correspondingly more plentiful and permanent ice is much less extensive than on the continental mainland.

The characteristics of Antarctic land life are the antithesis of land life in the tropics. Tropical plants and animals are often large, exotic and abundant in number of species. Antarctic plant life and Antarctic land animals are small in size and restricted in species. The reason is not far to seek; the higher plants at least require a basic minimum of soil, nourishment, sunshine, shelter from winds and frosts and above all a sufficiently prolonged period of summer warmth in which to reach maturity and bear fruit. The Antarctic lacks almost all these requirements. The centre of the continent is virtually a desert where only the indefatigable lichen finds a precarious hold on exposed rocks within 300 miles of the South Pole. Round the coast conditions permit the growth of a few mosses and algae. And it is only well to the north on the Antarctic Peninsula that one finds, in sheltered places, Antarctica's flowering plants, two grass species and a pearlwort. For a wider range of plant species you must go even further to

Iles Crozet, islands lying to the north of the Antarctic Convergence. Though virtually free of glaciers they have many Antarctic characteristics

South Georgia, lying in the colder waters to the south of the Antarctic Convergence, is free of permanent ice only in coastal regions

The white cliffs of the Antarctic Ice-Shelf, the seaward edge of the ice-sheet which envelops most of Antarctica

The mountains of the Antarctic Peninsula are sufficiently exposed to provide some of the continent's finest scenery

The Byrd Glacier, from 10,000 ft., among the highest of the many valley glaciers down which the ice from the polar plateau flows to sea-level

the north. South Georgia boasts coastal meadows of tussock-grass and several species of alpine plants. Nowhere south of the Antarctic Convergence is there any plant approaching even a shrub in size.

Where there are so few plants there are correspondingly few land animals; no land mammals, no reptiles, no amphibians are native to the Antarctic. The largest land animal in Antarctica is a wingless fly and most of the few remaining small creatures need to be viewed through a microscope. The islands have very little more to offer though South Georgia and Archipel de Kerguelen have reindeer and rats and Macquarie Island has rabbits – all imported by man.

In sharp contrast to the poverty of the land there is a richness of life dependent on the sea. There are, for example, some 40 species of birds breeding south of the Antarctic Convergence. These are mostly ocean feeders – petrels and penguins. Then there are a few gulls, cormorants, terns and sheathbills which get a living off the shore and in shallow waters; they are few in number and species. Only one land bird is recorded in the Antarctic – the South Georgia Pipit. We tend to think of the penguin as being the most typically Antarctic bird though there are species of penguin breeding as far north as the Equator. Of the six Antarctic breeding species only the Emperor Penguin winters on the continent, hatching its single egg on the floating 'fast ice' in the darkness of midwinter. The more common Adélie Penguin, whose inquisitive habits and comical waddle so endear it to visitors, migrates in the autumn to the pack ice. Penguins are, of course, flightless birds and therefore restricted to the coast. Not many birds go far inland. Only the beautiful white Snow Petrel and the Antarctic Skua are occasionally reported in mountain regions near the coast, though there are a few isolated reports of the latter within a hundred miles or so of the South Pole.

Frequenting the Southern Ocean are various species of whales and seals. Four of the seals are found mainly in the pack ice – they are the Ross, Weddell, Crabeater and Leopard Seals. Of these the Weddell Seal is the only coastal species. Away from the pack ice, on the more northerly islands of the Antarctic and sub-Antarctic, breed Fur Seals and Elephant Seals. Seasonal visitors to the Antarctic are the whales, marine mammals browsing with their young in the rich pastures of the Southern Ocean before returning to warmer tropical waters to the north. Whales found in the Antarctic vary much in size and shape from the vast Blue Whale, which weighs up to 150 tons and is the largest animal on Earth, down to the much smaller Killer Whales which hunt for seals and penguins in the off-shore waters of the continent. The Blue Whale, a valuable source of oil, blubber and bone, was until recently the staple of a prosperous Antarctic whaling industry. But, like seals in the past, whales in their turn have been grossly over-exploited with the result that a once profitable industry is now in full decline.

The South Pole – 90°S – the southern extremity of the Earth's axis, a point on the featureless Antarctic ice-sheet approximately 9,500 ft. above sea-level. Today it is the site of the U.S.A.'s Amundsen-Scott Station

Former Soviet scientific station Sovetskaya, surmounted by a bust of V. I. Lenin, marks the site of the Pole of Relative Inaccessibility, the region in Antarctica least accessible from the coast. It lies at lat. 82°06′S, long. 54°58′E on the high plateau of Greater Antarctica, 12,200 ft. above sea-level and 1300 miles from the coast

Paradoxical as it may seem, it is the cold polar seas which are the richest in potential nourishment for animals and perhaps, if tastes can adapt, for humans too. The Southern Ocean is an especially well-stocked larder. Upwelling currents distribute minerals which, with the light from the Sun, nourish a vast drifting community of tiny plants and animals known collectively as 'plankton'. This prolific marine life is the basic food of many fish, sea birds and certain whales. These animals in turn feed on each other and their decaying carcasses, by producing more nutrients help to keep a remarkable food chain in perpetual motion.

We now have a sufficiently general picture of the Antarctic regions before us to make a comparison with the opposite polar region, the Arctic, worthwhile. We tend to think of both as being rather similar. There are indeed likenesses; both have long, light summers and long, dark winters. Both are cold and characterized by large amounts of snow and ice. But only in a most superficial way are they truly alike. The differences are far more profound. The first of these was well epitomized by that great geographer Isaiah Bowman: 'the Arctic', he said, 'is a hollow and the Antarctic a hump'. As we have seen Antarctica is a lofty continent surrounded by an ocean; the Arctic is an ocean surrounded by the shores of the continents of North America, Asia and Europe. This essential difference has a fundamental effect on the climates of the two regions. The ice-covered Arctic Ocean, though comparable in area with the Antarctic continent, is a water mass whose warming effect moderates the climate of the surrounding region. Minimum temperatures in the central Arctic never plumb the extreme depths recorded in central Antarctica. Far lower temperatures are recorded in the Yukon Territory of Canada and in deepest Siberia than in the Arctic Ocean. The influence of the Arctic climate as it affects the northern circumpolar regions is very much less than that of Antarctica, whose greater-than-average height and vast store of ice generate a polar climate which can affect the growing crops on the far-away pampas of Argentina and chill the citizens of Sydney. The barren ice-clad island of South Georgia lies in latitude 54°S., 2000 miles from the South Pole. In the corresponding latitudes of the Northern Hemisphere lie flourishing centres of population like Liverpool and Belfast. To find a landscape similar to that of South Georgia in the Arctic you would need to visit the coast of east Greenland which is in latitude 73°N., roughly 1000 miles from the North Pole. Here in the brief weeks of summer Muskoxen graze among patches of small but brightly coloured flowering plants and a variety of animal life abounds including Arctic Fox, Arctic Hare and Polar Bear. The occasional settlement of Eskimo hunters is a reminder of the infinite adaptability of man to his environment. The same latitude in the Southern Hemisphere would present a very different scene, a waste of ice-shelves and glaciers, almost devoid of life.

Man in the Antarctic

The Antarctic regions have exercised the imagination of man since earliest times. It was the admiration which the Greeks of Classical Antiquity had for the perfect form that gave birth to the idea of a spherical Earth. It followed that the parts of a perfectly formed Earth must be in perfect balance and therefore the Greeks very logically postulated a frozen region in the South to balance the frozen Arctic of which they had some little knowledge. Serious exploration of the Antarctic has been in progress for something over 200 years, since 1738–9 in fact, when a Frenchman, Bouvet de Lozier, discovered the foggy ice-bound island known today by its Norwegian name of Bouvetøya. The Americans, the British and the French were among the early pioneers and they were subsequently joined by many other nations in the joint task of unveiling first the outermost islands and then Antarctica itself. The motives that impelled them were, as always in human affairs, very mixed. The early history of the sub-Antarctic islands is an ugly one, a systematic plundering of their only natural wealth, the Fur Seals and Elephant Seals, and sometimes when these were scarce, the penguins were callously boiled alive for their oil. But discovery of the islands was to lead to discovery of the continent and here, since there were few natural riches to whet his baser appetites, man could indulge that nobler side of his nature which seeks knowledge for its own sake. All the great Antarctic expeditions since Captain Cook's first circumnavigation of the continent in 1772–5 have had a strong scientific element in them, but science often had to take second place to national interests or individual whims. Antarctic exploration suffered, too, from a lack of continuity and a lack of co-ordination between the different countries which took part. This led to much duplication of effort in some fields and complete ignorance in others. Consequently, our knowledge of the Antarctic grew very slowly until the advent, little more than a decade ago, of the planned international collaboration that we see today. As forward-looking scientists had been aware for very many years, the large-scale investigation of Antarctica could only be brought about by a pooling of resources. This could only happen when the international and political climate was favourable and when governments were liberal enough to provide the large sums of money which modern Antarctic exploration, with its costly scientific programmes and even more costly supporting equipment, requires. By the end of the Second World War technology had advanced to a point when it could be harnessed to virtually any problem relating to the Earth and outer space, however vast the scale. The first mass scientific invasion of the Antarctic was launched to support the International Geophysical Year of 1957–8, a world-wide programme of co-ordinated research in the Earth sciences – oceanography, meteorology, upper-air physics and glaciology. Particular attention was focused on the Antarctic, where scientific observers from

The tellurometer – a highly accurate instrument for measuring distance in inhospitable country. The system is based on the use of two portable microwave transmitter-receivers

12 nations set up stations not only on the islands but round the shores of Antarctica and – for the first time – at the South Pole itself. So successful was the outcome of this international scientific effort that it was decided to carry it on indefinitely. In 1959 an Antarctic Treaty was signed by the co-operating nations at Washington, D.C., U.S.A., which reserved Antarctica and its surrounding ocean and islands south of latitude 60°S. for peaceful scientific purposes by all interested countries. The signatories included such ideologically opposed nations as the U.S.A. and the U.S.S.R. All military activity was banned and political claims were 'shelved' for 30 years. Now, for the foreseeable future at least, the Antarctic is a region reserved for science. Man has come to stay and an era of permanent occupation has succeeded the old pattern of spasmodic expeditions.

The Antarctic regions can now be said to boast a permanent population. Some 800 scientists – the number fluctuates – annually winter on the Antarctic continent. In the summer, when the relief ships arrive, the number temporarily swells to perhaps 2000. But there are few places in the Antarctic, or the sub-Antarctic, where you will find settled communities, such as we know them, where normal family life can be enjoyed. The Falkland Islands have around 2000 settlers and South Georgia has a small population of administrators and scientists. Only two women are recorded as having wintered south of the Antarctic Circle and there is a general reluctance to admit them to what is still essentially a man's world. Of the 37-odd stations presently manned in the Antarctic and sub-Antarctic none, excepting Stanley in the Falkland Islands, ranks higher than a small hamlet. The largest station in the Antarctic is the United States' McMurdo Station, a sprawling conglomeration of huts on Ross Island which has all the makeshift qualities of a Yukon frontier town in the heyday of the Gold Rush. Nevertheless, McMurdo, with a population of over 700 in summer and 200 in winter, can claim many of the more sought-after amenities of civilized life – a church, a shop, a cinema and a gymnasium, in addition to a centrally heated barracks with well-equipped cafeterias offering a variety of excellent dishes. Power for the station comes from a nuclear reactor and a desalination plant provides fresh water from the sea. Life in the field is correspondingly cushioned against the forces of nature. A whole station, including huts, scientific equipment, stores and vehicles, can be ferried from one part of Antarctica to another by air transport. A wide range of vehicles – motor toboggans, tracked vehicles of many kinds and helicopters – can take the scientist to his place of work with the minimum waste of time and the maximum degree of comfort. Scientifically designed clothing and properly balanced rations will protect him from those two bugbears of polar exploration – frostbite and scurvy. This is the accepted way of life in the Antarctic today; there is no virtue in discomfort for its own sake. Perhaps, too, these high living standards are a pointer to the future when pressure for living space in the world's more crowded latitudes may enhance the value of Antarctic real estate.

Boundaries and place-names

We have discussed the Antarctic as a geographical entity. But what of its political status? Who owns the Antarctic continent and the surrounding islands? During the course of its history a number of nations have staked out claims based on priority of discovery and continuity of occupation and administration. As we have seen, the Antarctic Treaty of 1959 put all claims to territory lying south of latitude 60°S. in cold storage for 30 years. In consequence (except in Latin America) Antarctic territorial claims now have a faintly musty smell, and it is to be hoped that with the passage of time they will be quietly forgotten. Current maps of Antarctica rarely show the cake-like sectors radiating from the South Pole into which the Antarctic continent is divided. Nevertheless, we cannot completely ignore these divisions nor can we pretend that their political names do not exist. The claimant nations, as the lawyers term them, still reserve their rights over their respective Antarctic sectors, administer justice, operate post offices, allocate place-names and exercise in various ways their sovereign functions. Thus the British Antarctic Territory, which includes the whole of the Antarctic Peninsula, is regarded as a colony under the British Crown administered by a High Commissioner. Argentina and Chile also have overlapping claims in this region which in turn overlap those made by Britain. Other Antarctic 'powers' are New Zealand (Ross Dependency), Australia (Australian Antarctic Territory), France (Terre Adélie) and Norway (Dronning Maud Land). One large sector, Byrd Land, has never been claimed at all. Historically it has always been associated with United States' activity. The United States, however, neither claims territory on the Antarctic continent nor recognizes the claims of other nations. The U.S.S.R. is another active nation in Antarctica which neither makes nor recognizes territorial claims. Its sphere of activity during the past decade has been in the Australian Antarctic Territory as the many Russian place-names in that area testify. The islands north of the latitude 60°S. treaty line are administered by Australia, France, the United Kingdom, New Zealand, Norway and South Africa. Two British islands to the north of this line – South Georgia and the South Sandwich Islands – are dependencies of the Falkland Islands.

The lack of international agreement on the use of geographical names is a world problem of which the Antarctic provides some of the most glaring examples. Argentina and Chile, for example, who dispute British sovereignty over what today is unequivocally named the Antarctic Peninsula, were wont to describe the latter as respectively, Tierra San Martín and Tierra O'Higgins. The U.S.A., a non-claimant nation, but asserting priority of discovery, named it Palmer Peninsula. The British themselves called it Graham Land. The task of the map-maker was, in consequence, not an easy one and though nowadays there is a great deal more collaboration between national place-naming authorities to avoid the allocation of different names to the same features, and though it is usually possible to obtain agreement on names between English-speaking nations, there can never be complete international agreement on such matters because of differences in language. Throughout this book we shall be using the place-names authorized by the administrative authorities of each area as recognized by the United Kingdom; in consequence we prefer Dronning Maud Land to Queen Maud Land, Terre Adélie to Adélie Land, Bouvetøya to Bouvet Island and so on.

Nearly 200 years have passed since Captain Cook, having circumnavigated the world in high southern latitudes without sighting an Antarctic continent, commented in his journal: '. . . if anyone should have the resolution and perseverance to clear up this point, by proceeding further than I have done I shall not envy him the honour of the discovery; but I will be bold to say, that the world will not be profited by it'. On the first count history has shown Cook's prognosis to be well off the mark. But what of the second? To what extent can the world benefit from Antarctic science and exploration? This is a difficult question to answer; the chapters that follow may help the reader to reach his own conclusions.

Economically the Antarctic has little to offer now. Whaling, as we have seen, is on the decline and sealing is not likely to be resumed on a commercial basis until much-depleted stocks are fully restored. The Southern Ocean is indeed rich in raw proteins – but not, unfortunately, in marketable fish. Mineral wealth Antarctica assuredly has in plenty – coal and uranium deposits have been reported – but this is for the most part hidden and inaccessible under the all-enveloping ice sheet. Tourism has recently started in a small way and may expand measurably in the years to come. Antarctica may well have an economic future, but it seems unlikely to follow orthodox lines. This leaves scientific research as the Antarctic's main industry with knowledge its main export and the benefits are likely to be long-term rather than immediate. Nevertheless, 12 nations are currently spending large sums of money on scientific programmes in the area. The total annual bill for the U.S.A. alone is in the region of $30 million and the United Kingdom on a smaller but no less effective scale is spending about £1 million annually. These sums suggest that supporting governments are convinced of the worthwhileness of the end-product. Let us briefly review just a few of the possible benefits of Antarctic science.

We are all aware of the need for a reliable long-term weather forecasting system. The future welfare of a growing world population may hang on a solution to the problem; governments look to climatologists for long-term solutions to problems of water supply for irrigating world deserts and providing for the needs of populous conurbations. The answers will be found in a much fuller understanding of world weather patterns and their mechanism than we have at present. Antarctica, as we shall see in another chapter, plays a significant part in the global weather pattern. An understanding of the principles that govern the Antarctic weather system will perhaps bring the meteorologists measureably nearer a solution to these puzzles. Closely linked to the problems of climate is the nature and growth of the Antarctic ice-sheet – is it shrinking, increasing or simply static? We need to have a long-term answer. For locked up in all this ice is a frozen reservoir of fresh water equal in volume to the entire North Atlantic Ocean. Were it all to melt the world's sea-levels would rise perhaps 200 ft. inundating all low-lying coastal regions, including our major ports and centres of population. We are all very much aware these days of the problems of outer space, not only those concerned with space travel but those that affect our own atmosphere. Here, too, Antarctica provides an ideal vantage-point for space observation. For it is in the polar regions, where the Earth's magnetic lines of force enter the atmosphere, that one can most conveniently study the effect of the Sun's radiation on the region known as the 'ionosphere', and note its reaction with the Earth's own magnetic field. Undue disturbances in this layer of the upper atmosphere can cause complete failure of short-wave radio communication, sometimes on a world-wide scale, and give rise to the strange and beautiful phenomenon called 'aurora'. From other Antarctic studies we learn much too; Antarctic geology helps us to understand the relationship of Antarctica to the surrounding continents now and in the very remote past; its fossils are indicators of ancient fauna and ancient climates. In short, all Antarctic science can be justified inasmuch as we cannot pretend to understand this Earth of ours without a detailed knowledge of the polar regions which form such a sizeable area of its surface. To expect immediate practical results is unrealistic and unfair to science. Clothing and feeding men at low temperatures; the design of portable huts; operating and servicing aircraft and ground transport in the most testing of all nature's environments; the solution of these problems brings practical advantages to us all. Perhaps the most convincing justification for the men, the materials and the money that go into Antarctic research is to be found in the political advantages. For the Antarctic Treaty exemplifies, as no other treaty in history has done before, the part that science, mankind's common language, can play in promoting international understanding, reducing international tension and furthering the cause of world peace.

Tourism promises to be an important Antarctic industry. The following pictures were taken on an expedition to the Antarctic Peninsula organized by Lindblad Travel, New York:

A landing craft bearing tourists towards their first Antarctic landfall

Weddell Seal-spotting on Half Moon Island

Landing on Anvers Island

Paradise Bay, cruise ship in the background

2
THE SNOW AND ICE

In our opening chapter we saw that the Antarctic continent and many of the surrounding islands are characterized by perpetual snow and ice. In this chapter we shall be looking at Antarctica mainly from the standpoint of a glaciologist, glaciology being the study of snow and ice in all its forms. We shall be considering not only the ice which covers the Southern Ocean for the greater part of the year, but also the ice which entombs much of the landscape of the Antarctic continent.

Ice and its physical properties

But before considering ice in its natural state let us firstly review some of its general properties. When water freezes it turns into the crystalline solid with which we are all familiar, namely ice. Water because of its sheer universal presence has for long served as a physical standard. In thermometers its freezing-point is used to determine zero in the Réaumur and Centigrade scales, 32° in the Fahrenheit scale. Water and ice display certain characteristics which are to some extent unique in nature. On freezing, pure water expands by nearly 9 per cent of its volume, a textbook fact demonstrated annually in wintertime by split water-pipes and burst car radiators. At 0 °C (32 °F) the density of ice is only 0.917 that of water which explains why in cold weather ice forming on ponds floats on the surface. If the ice were denser it would sink to the bottom and, since the surface would continue to cool rapidly, the pond would freeze from the bottom upwards until it became solid. We say that the melting-point of ice is 0 °C but this is only true under atmospheric pressure at sea-level. If the pressure increases the melting-point of ice falls by 0.0075 °C for one atmosphere increase in pressure. It is this property of ice which enables one to skate on the frozen surface of a pond; the pressure of the skate melts the surface of the ice and the thin film of water resulting acts as a lubricant. The same property of ice plays a part in the movement of glaciers over their beds.

The thermal properties of water and ice are also of significance because of their effect on climate. In order to melt ice at 0 °C into water at the same temperature a certain quantity of heat must be supplied namely 80 calories* per gramme. Conversely when water freezes the same amount of heat has to be dissipated. This property of water and ice to radiate and absorb large quantities of heat has a powerful ballasting action on extreme variations of temperature. For example, extremes of temperature on the islands fringing Antarctica are much less excessive than the extremes of temperature experienced on the high plateau 1000 miles inland.

Ice has numerous other properties, an understanding of which enables glaciologists to predict its behaviour in the natural state and engineers to put these properties to practical advantage. Ice is elastic, which means that it has a certain 'give'; it is also plastic which means that it can be permanently deformed without breaking due to slow continuous deformation when subjected to pressures greater than a certain value. It is important to understand these properties whether one is studying the flow of glaciers or planning to build a scientific station or an air-strip on the Antarctic ice-sheet.

In nature ice will be found in many different forms; in the Antarctic we shall consider firstly the ice found floating on the surface of the ocean and then the ice which covers the land.

Floating ice

The first hint of ice as one approaches Antarctica by sea will be a white glare on the distant horizon, known to mariners as *iceblink*, which is the light reflected from the approaching pack ice. *Pack ice* is the term given to fields of ice that drift with winds and currents. Its navigability is measured in tenths; thus 10/10ths, or very close pack, means a solid mass with very little water visible; 1/10th to 3/10ths would indicate very open pack with much open water. Nearer to the coast of the continent sea ice becomes attached to the shore or maybe to shoals when it is known as *fast ice*. Fast ice may extend from a few feet to several miles from the coast. When its surface exceeds more than 6 feet above sea-level it is known as an *ice-shelf*. Ice-shelves, as we shall see later on, are the source of many of the huge icebergs to be found in the Southern Ocean.

Sea ice is different in structure to ice formed from fresh water because of its salt content. The world's oceans are salt because rain falling on the land washes out soluble chemicals from the earth and rocks and carries them out to sea in solution. As the sea water evaporates the dissolved salts remain behind. Their distribution is fairly uniform though it varies somewhat from ocean to ocean. Sea water of average salt content freezes at about −1.9 °C. When this happens small crystals of ice, usually square or

* A calorie is a unit of heat, namely the amount of heat required to raise the temperature of 1 gramme of water by 1 degree Centigrade.

Open pack ice consisting mainly of small floes. The navigable passage is known as a *lead*

hexagonal in shape, form just below the surface of the water. These platelets, or *frazil ice*, form a surface slush with a slightly oily appearance, a hint to the captain of any vessel lingering in high southern latitudes that he should be making for home. The action of wind and waves will tend to compact the plates together forming a continuous cover of ice called *grease ice* which has a matt appearance. This thick soupy layer of ice is remarkably flexible, waves from the wake of a ship can travel through it without breaking it up. As freezing proceeds the grease ice forms a thin, elastic shiny crust of ice which in turn, by the action of wind and wave, breaks up into separate masses called *pancake ice*. These plates of ice, characteristically circular in shape and measuring between 2 to 6 ft. in diameter, tend to acquire turned-up edges through colliding with one another which give them the appearance of enormous water-lilies. These finally freeze together forming level sheets of ice from 2 to 6 in. in thickness.

Icebreakers U.S.S. *Edisto* and U.S.C.G.S. *Northwind* heading R.R.S. *John Biscoe* in convoy through young ice prior to entering the pack

Below: Pancake ice. The 'water-lily' effect is caused by the pieces of ice colliding with one another

During the Antarctic winter this *new ice* continues to grow up to 6 ft. or more in thickness in its first two years of existence and may reach 9 ft. or more in thickness if it survives even longer. *Old ice* is an apt enough term for this variety of sea ice; it is recognizable by its blue colour and is usually rougher in appearance than young ice. At this stage of its growth most of the salt will have been leached out so that it forms a readily available source of drinking-water for Antarctic expeditions.

The action of winds and currents will keep floating pack ice in a constant state of motion. A violent storm may break it up forcing floe against floe and causing *hummocking*, that is the haphazard piling of one piece of ice upon another to heights of 30 ft. or more. The overriding of one floe by another is technically termed *rafting*. The term *floe* is reserved for any piece of floating pack ice exceeding 30 ft. across. Smaller pieces of floe are referred to as *ice-cakes*. The smallest bits of all, the wreckage of all the other forms, are collectively called *brash ice* and these are never more than 6 ft. across.

We have already referred to that other kind of sea ice, afloat but not freely floating, called *fast ice*. In winter this ice may be firmly fixed to the shore, but in summer, assisted by the rising and falling of the tide, it may break away and float out to sea leaving behind a narrow strip of ice attached to the coast called an *ice-foot*. The junction between the ice-foot, which remains unmoved by tides, and the fast ice, which rises and falls with them, is known as a *tide crack*.

Tide crack at Vassfjellet, Dronning Maud Land

Icebergs

Not all the floating ice of the Southern Ocean is composed of salt-water ice; the giant tabular icebergs, which are so characteristic of Antarctic waters, are derived from the ice-shelves and glaciers of the Antarctic continent which are, of course, formed largely from falling snow. Tabular icebergs are flat-topped and rectangular in shape and are of a peculiar white colour and lustre like plaster of Paris, due to their relatively large air content. The commonest exceed a mile in length but outsize bergs are constantly being reported. One such measured 90 miles long and between 100 and 130 ft. high. Occasionally icebergs with a more irregular surface may be seen, dazzling white in colour and relieved perhaps by hues of green or blue, with occasional characteristic bands of sand or debris streaking the sides. The glacier bergs are derived from glacier tongues flowing directly into the sea. Occasionally icebergs become stranded on submarine shoals and may accumulate a dome of snow before finally floating free again. Smaller, irregularly shaped icebergs, often weathered by the action of the sea into grotesque pinnacled shapes and reminiscent of the icebergs typical of Arctic waters, are formed from the break-up and decay of larger icebergs. It is often said that nine-tenths of an iceberg are beneath the surface of the sea; if icebergs were to consist of pure ice this would be so but Antarctic icebergs contain vast quantities of air-bubbles so that perhaps only four-fifths of their bulk is normally submerged. Because most of an iceberg is underwater its movement is controlled by

A new iceberg is calved as a huge section of the ice-shelf breaks away and drifts off to sea

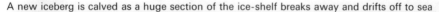

A tabular iceberg. Typical of Antarctic waters these bergs can exceed 90 miles in length and 100 ft. in height above sea-level

currents rather than by wind. Once clear of the pack and out into warmer water icebergs tend to disintegrate fairly rapidly by weathering. Salt water dissolves the underwater surface until the smaller bergs capsize.

Extent of floating ice

During the summer months Antarctica is surrounded by a broken fringe of pack ice which may be separated from the coast by a belt of water a few miles wide. In the Ross Sea it is not uncommon for the ice to disappear almost entirely. As winter approaches the belt of intervening water, called a *shore lead*, will begin to freeze and join the pack-ice fringe to the coast. Violent winter storms may often open up the shore lead. The maximum extent of the pack occurs in August or September when its average northern boundary will be latitude 54°S. in the Atlantic sector, latitude 56°–59°S. in the Indian Ocean sector and latitude 60°–63°S. in the Pacific sector. During a severe winter this ice boundary can be extended from 2° to 9° of latitude northward. In coastal waters the regional sequence is as follows. In early winter, that is towards the end of February, ice will form in the southerly coastal regions of the Weddell Sea. Shortly after this ice begins to form in the Bellingshausen Sea. Lastly, in March or April, the Ross Sea starts to freeze. Melting occurs in the early summer in the north and

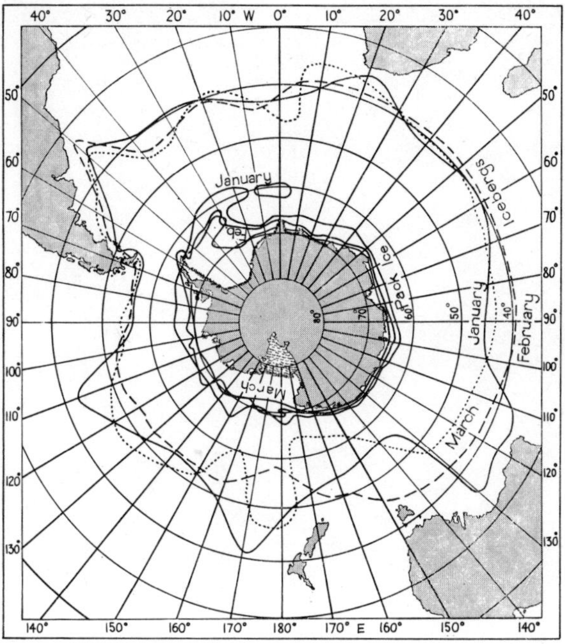

Map of the Antarctic showing extreme limit of icebergs and mean limit of pack ice in summer. *(From 'The Antarctic Pilot', 3rd Edition 1961)*

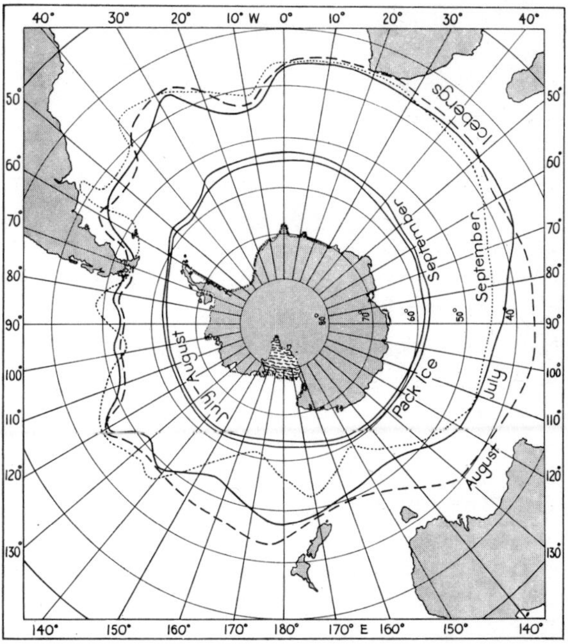

Map of the Antarctic showing extreme limit of icebergs and mean limit of pack ice in winter. *(From 'The Antarctic Pilot', 3rd Edition 1961)*

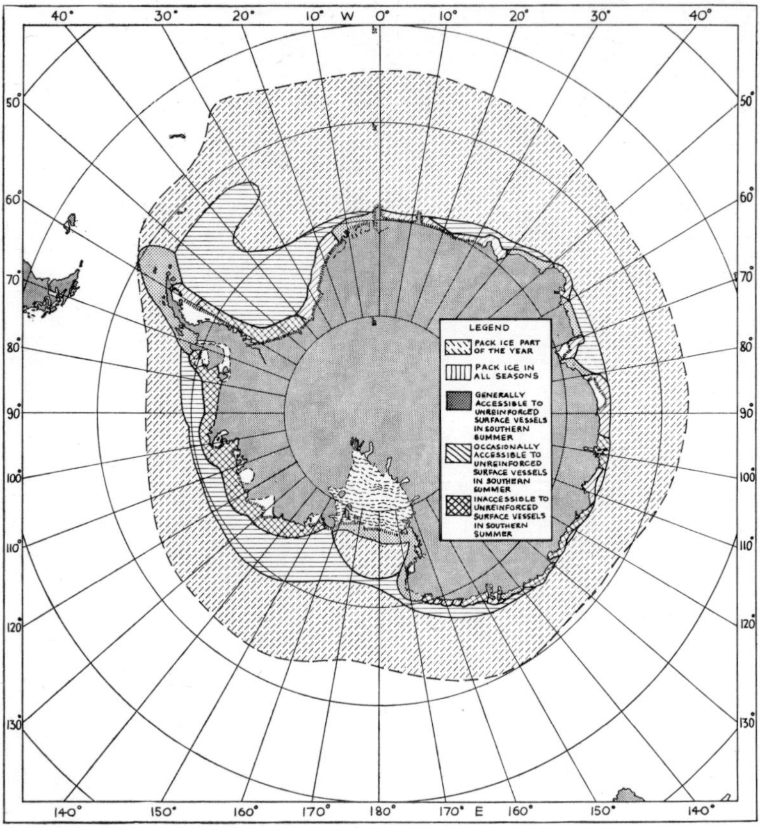

Map showing accessibility of the coast of Antarctica. *(From 'The Antarctic Pilot', 3rd Edition 1961)*

spreads gradually southward so that by January the belt of pack ice is broken up leaving only the Weddell and Bellingshausen Seas with large masses of sea ice in them.

Movement of floating ice

Antarctic pack ice is never entirely still; winds and currents keep it constantly on the move though the rate of movement tends to be less in winter than in summer. The direction of movement is generally from east to west propelled by the prevailing easterly winds. Measurements taken on ships beset in the Weddell Sea indicate that in these waters the drift of the pack is in a clockwise direction, resulting in a west to east drift in the northern reaches. In the Bellingshausen Sea it appears that the prevailing east to west drift is less strong than elsewhere and this may account for the fact that the build up of ice in this area is greater than anywhere else in the Southern Ocean.

Ice prediction

The movement of shipping to and from Antarctica is not a major problem though obviously ships must be properly strengthened for use in ice and their captains experienced in the techniques of ice navigation. Otherwise sufficient is known about the seasonal variations of the ice to ensure that ships get through to their bases almost as a matter of routine. But from time to time the unexpected does happen. One year there may be no ice at all in a particular region; the classic example is James Weddell's remarkable voyage in 1823 when he penetrated as far south as latitude 74° 15′S., longitude 34° 16′W. Another year there may be more ice than normal and then a ship can find itself beset and the personnel of a station may go unrelieved. Though limited studies have been made on Antarctic sea-ice distribution, based on historical records, no ice-prediction service yet exists comparable with that available in the Arctic and adjacent oceans. Such a service would be a great boon to the navigators of ice-breakers but unfortunately the small number of meteorological stations and the almost complete absence of weather-ships in Antarctic waters has made prediction difficult up till now. However, the development of the meteorological satellites Nimbus I and II has revolutionized the technique of forecasting. The Nimbus satellite, which is in continuous polar orbit, can take visual and infra-red photographs of the pack ice over a wide area from a height of some 500 miles. These photographs show the cloud cover or ice boundary round the whole of Antarctica and with experience one can soon distinguish between clouds and ice. Synoptic data from these satellites over a long period of time is likely to provide a basis for a reliable means of forecasting the distribution of the Southern Ocean pack.

Land ice

The history of how the land ice of Antarctica is formed and how it finds its way into the sea is a fascinating one. Land ice, by which we mean ice lying on solid land as distinct from the ice formed by the freezing of rivers, lakes and ponds, is for the most part derived from ice formed from fallen snow. At temperatures above freezing-point the atmosphere always contains a certain amount of moisture in the form of invisible water vapour. The warmer the air the more moisture it can carry; conversely as it becomes colder a point is reached when the air can carry no more and the water vapour condenses. At temperatures below freezing ice may form either as pellets or in the more regular form of ice crystals. When both the temperature and the water vapour content of the air are low the crystals often take the shape of single plates or columns. Such is the form of snow found on the high mountains and plateaux of Antarctica. At temperatures lower than −40 °C. the size of these plates may be less than a millimetre in diameter and they have the consistency of fine flour. At temperatures higher than −5 °C. the crystals agglomerate into the familiar

Snow consists of ice crystals made of water vapour condensed on a nucleus by sublimation at low temperatures. Crystals usually have the shape of a hexagonal plane or column and almost innumerable variations in form. W. A. Bentley, from whose book *Snow Crystals* these are reproduced, photographed at least 6000 of them snowflakes. The light reflected from the crystal facets gives the snow its familiar white appearance. The basic snow crystal is hexagonal in form. Complex and beautiful patterns, branched like trees, can be created by various forms of snowflake growth. Over 4000 different forms of snowflake have been photographed and these may represent only a fraction of all the possible types.

How glacier ice is formed

In Antarctica snowfall is not heavy, averaging 6.5 in. of equivalent water a year. Losses of snow by melting are negligible because of the low temperature. Snow has piled up slowly over Antarctica for so many thousands of years that it now covers 95 per cent of the continent with an ice-sheet whose weight is sufficient to depress part of the underlying land mass well below sea-level. The process by which snow turns into ice is complex. It is analogous to the metallurgical technique known as 'sintering' whereby a metallic powder is transformed into a solid mass by the application of heat and pressure. It could be matched with the geological consolidation of sand into sandstone or clay into shale. Essentially it is a process of snow densification involving one or more different mechanisms. On cold ice-sheets like those of Antarctica and Greenland, where there is little melting and refreezing on the surface, the transition is essentially a dry one. What happens is in very general terms as follows. The snow as it reaches the surface of the ice-sheet consists of hexagonal crystals of varying forms. Within a day or two the crystal shapes, due in part to local evaporation and condensation, become rounded forming a loose aggregate of spherical grains which, because of the intervening air spaces, is porous to air. With increasing depth and pressure the grains become more and more closely packed and the air spaces separating them correspondingly smaller. Eventually a point is reached when no more packing is possible and a state of minimum porosity is reached. Further application of stress with increasing depth brings a fresh mechanism into play involving a change in size and shape of the grains which tend to join together. The air spaces, in consequence, gradually become more isolated from one another and less permeable to air. During this stage, between maximum densification by packing to the point where all the air spaces are sealed off, the material is known as *névé* (or sometimes as *consolidated snow* or *firn*). It is this change from a permeable to a non-permeable material, taking place gradually over a period measured in tens or even hundreds of years, that marks the change from névé to *glacier ice*.

Approximately 95 per cent of Antarctica's ice-sheet is formed of glacier ice of this kind. The process of change from snow to ice affects not only the nature of the original snow but also its appearance. Firn is less white than snow and as it approaches the glacier-ice stage it becomes more translucent, frequently with a blue or green colour when seen in deep

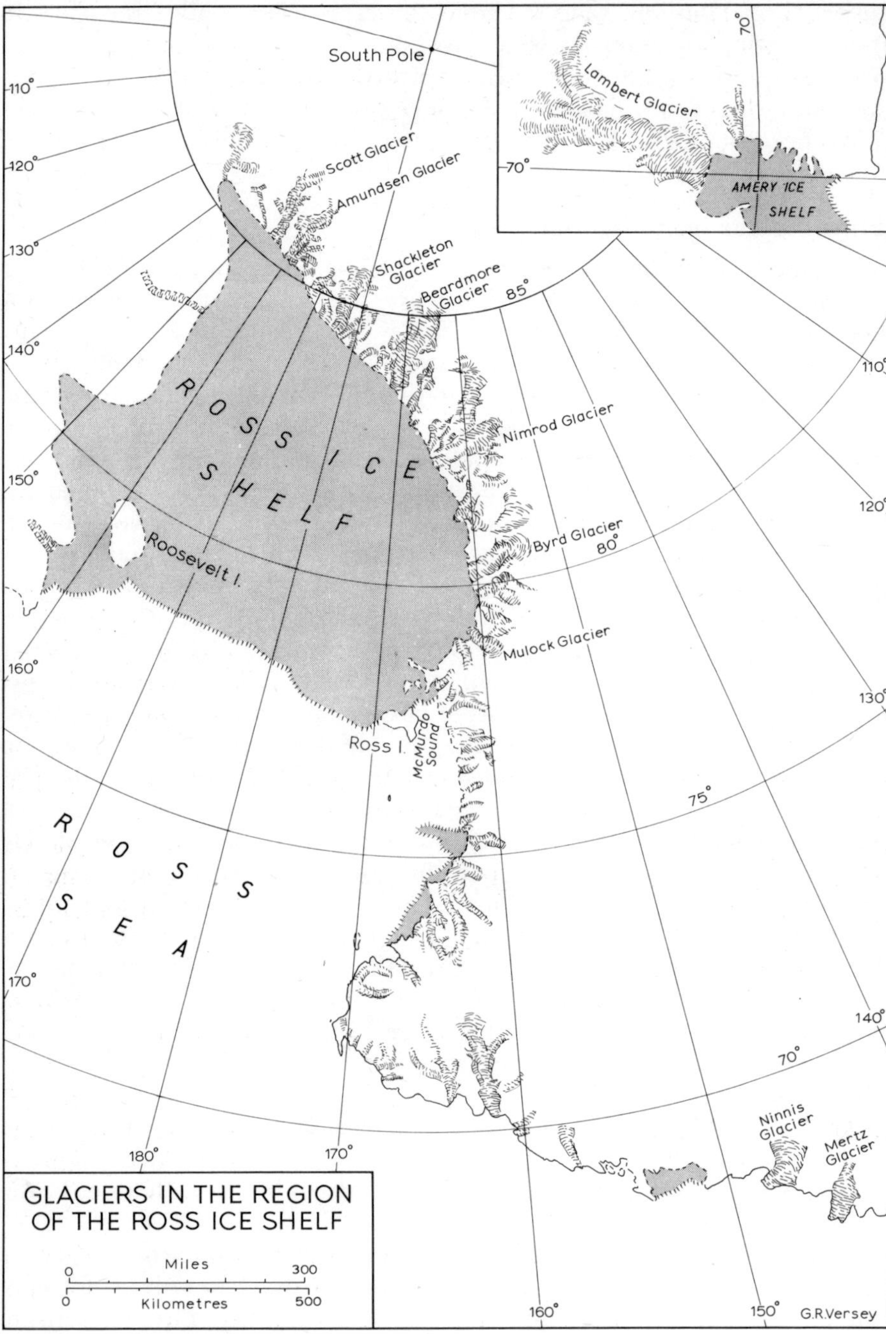

GLACIERS IN THE REGION OF THE ROSS ICE SHELF

crevasses. The depth at which the transition from firn to ice takes place depends on such factors as the temperature and the rate of snow accumulation at the surface. In Antarctica this transition occurs at an increasing depth as one travels inland from the coast. At the South Pole, where temperatures are low and snowfall slight, glacier ice is formed at about 300 ft. in snow which is about 1000 years old. On the Ross Ice-Shelf, nearer the coast, where temperatures and snowfall are higher, the transition takes place at between 120 and 200 ft. in snow between 200 and 300 years old.

Movement of the Antarctic ice-sheet

One might suppose this vast mass of accumulated ice to be a static immovable lump; on the contrary it is in a state of constant motion, a fact partly explained by the viscous and plastic properties of ice which enable it to flow and to bend under pressure. Large bodies of ice, such as the ice-sheet of Antarctica, are able to flow outwards under their own weight as a result of acceleration due to gravity. The rate at which they flow will depend on a number of factors such as the dimension of the body of ice, its thickness and the slope of its surface. It is the surface, rather than the bedrock slope which is important, for much the same reasons that the surface slope of a river or lake, rather than the bottom, causes it to flow. A further important factor determining flow is the temperature of the ice, especially on the bed. The mechanism by which glaciers are enabled to

The Antarctic ice-sheet photographed from a height of 1000 ft. showing sastrugi a foot or more high

Glacier tongue, Ross Island – the floating, seaward projection of a glacier

Scott Glacier – a valley glacier discharging into the Ross Ice-Shelf

Commonwealth Glacier, McMurdo Sound – the lobe-shaped terminus spread out over lowlands at the base of mountains is known as a *piedmont glacier*

The Taylor (*centre*) and Ferrar (*top right*) Glaciers of Victoria Land drain the inland ice-sheet through gaps in the peripheral mountains. Valley glaciers of this kind are known as *outlet glaciers*

An air photograph of the Axel Heiberg Glacier, Victoria Land, from 25,000 ft., showing the route to the polar plateau (*top background*) taken by Amundsen on his dash to the South Pole in 1911

flow is one that has puzzled glaciologists for many years. Undoubtedly the process is a complex one which can involve both the sliding of the ice-sheet over its bed as well as creep within the ice itself, a process of gliding caused by deformation and changes of the ice crystals. It is unknown how much the Antarctic ice moves by sliding on its bed as compared with the process of internal creep. As to the rate of flow of the ice-sheet this is probably a few yards a year only over much of the inland parts of the ice-sheet. One estimate reckons that it would take a particle of ice deposited at the Pole of Inaccessibility over 100,000 years to travel to the edge of the ice-sheet.

Physical features of the Antarctic ice-sheet

The Antarctic ice-sheet consists of two main regions, those of Greater Antarctica and Lesser Antarctica. The ice covering these regions was itself derived from the coalescence of a number of smaller *ice-caps.** The ice of Antarctica's interior is referred to as the *inland ice-sheet*; it rests on the underlying rock and is characterized by its great thickness, over 6000 ft. on average and attaining 12,000 ft. in places, sufficient to mask most of the underlying topography. The general slope of the inland ice-sheet is everywhere towards the coast so that the ice flow is outwards and downwards. As the ice nears the coast its velocity increases and its thickness decreases. Eventually it reaches the sea either as a broad *ice-front* or *ice-wall*, as in many places along the coast of Greater Antarctica. Alternatively, it may emerge in the form of *valley glaciers*, *ice-streams* or *ice-shelves*.

Valley glaciers

Where the inland ice begins to thin out, especially near the coast, nunataks or even whole mountain ranges may protrude through the ice channelling it into relatively fast-flowing valley glaciers; in the mountain ranges of Victoria Land, southward towards the Queen Maud Range are some of the most spectacular glaciers in the world, similar to, but larger than, those of the European Alps. Seven major valley glaciers drain into the Ross Ice-Shelf – The Mulock, Byrd, Nimrod, Beardmore, Shackleton, Amundsen and Scott Glaciers. Enshrined in history, though not the largest, is the Beardmore Glacier, 100 miles long and 12 miles wide on average, down which struggled Captain Scott's ill-fated party on their historic return journey from the South Pole in 1912. A recent survey of these glaciers shows that they are moving at an average speed of 1100 ft. a year and are adding some 15 square miles annually to the ice-shelf. On the other side of the continent, in Mac.Robertson Land, feeding the Amery Ice-Shelf, is the Lambert Glacier, probably the world's largest valley glacier.

* An ice-cap is a dome-shaped glacier usually covering a highland area. Ice-caps are considerably smaller in extent than ice-sheets.

Ice-streams and glacier tongues

In some places rivers of ice debouch into the sea whose margins may or may not be clearly defined. These *ice-streams* are a part of the main ice-sheet which flow more rapidly than the surrounding ice. The rate of flow of many Antarctic glaciers and ice-streams is sufficient to carry them some distance out to sea as floating *glacier tongues*.

Ice-shelves

More than 10 per cent of the Antarctic ice-sheet does not terminate at the coast but extends out to sea as floating *ice-shelves* reckoned to form one-third of the Antarctic coastline. As we have seen these ice-shelves are fed to some extent by the fast-moving glaciers to landward, but the chief source of their nourishment is snow accumulation on their upper surface. They are often prevented from breaking away to form icebergs by restraining features such as the headlands of deep bays, drowned off-shore islands and shoals; without these restraints they would tend to float out to sea and break up. The calving of the forward edge of an ice-shelf produces the giant tabular icebergs typical of Antarctic waters. Frequently, at the boundary dividing an ice-shelf from the mainland or the parent land ice are found parallel lines of *strand cracks*; characteristic features of this zone, these cracks, about half an inch in width with raised rims, open and close as the tide bends the ice junction between inland and shelf ice. The largest ice-shelves are the Ross and Ronne Ice-Shelves. The Ross Ice-Shelf, discovered by Sir James Clark Ross in 1841, covers an area of some 310,000 square miles, a region larger than France. It varies in thickness from about

The *ice front* or seaward edge of the Antarctic Ice-Shelf

The edge of the Ross Ice-Shelf from 4400 ft. The light area in the sea (*right*) is newly formed ice. The wide fissures to the left of the ice-shelf are crevasses

600 ft. at its seaward edge to 2000 ft. at its junction with the inland ice-sheet. The Ronne Ice-Shelf, occupying an embayment in the Weddell Sea to the west of the smaller Filchner Ice-Shelf, has an area of 127,380 square miles. It is named after the American, Finn Ronne, who discovered it in 1947. Smaller ice-shelves include the Getz and Larsen Ice-Shelves in Lesser Antarctica and the Lazarev, Shackleton, West, Amery and McMurdo Ice-Shelves in Greater Antarctica.

Ice surface features

As a result of the numerous elevation measurements made on the inland ice-sheet during the past ten years, we now know that the surface is far from being the smooth featureless dome it was once envisaged to be. Greater Antarctica turns out to be an irregularly shaped elliptical dome slightly off-centred on the Pole of Inaccessibility and with a maximum height above sea-level of some 13,000 ft. On the north coast of Greater Antarctica the regularity of the dome is dented by the Lambert Glacier and the Amery Ice-Shelf. In Lesser Antarctica the ice-sheet formation is more complex consisting of two imperfect domes, the larger of the two forming a divide between the ice flowing into the Ross Ice-Shelf and that flowing into the Amundsen Sea; here the maximum elevation is only 6500 ft.

On a smaller scale the ice-sheet is moulded into numerous forms such as the great terraced steps of Edward VII Peninsula or the wave-like undulations observed by the Commonwealth Trans-Antarctic Expedition in 1957–8. Features such as these are most probably reflections of the landscape buried many thousands of feet below the ice surface; certainly they are often very subdued and are much smoother than the irregular bedrock beneath. Other features are often more conspicuous from the air than from ground-level. Such are *crevasses*, great cracks in the ice that form at places where the ice is stretched to breaking-point because of changes of direction, flow or surface slope. Some of the most heavily crevassed areas in Antarctica are to be found in regions where ice-shelves meet land ice, or in the fastest flowing valley glaciers. On the inland ice-sheet crevasses are less frequent except in the regions of nunataks or where submerged mountain-peaks approach the surface. A crevassed area on the inland ice-sheet may extend for several miles with fissures well over 100 ft. deep and 50 ft. or more across. Smaller crevasses are often invisible from the ground because of the layers of snow which frequently cover them. Larger crevasses may be covered by a thick snow bridge strong enough to support a man with a team of sledge dogs or even a heavy powered vehicle, but such crevasses represent potential disaster, especially to heavier vehicles.

Other small-scale surface features on the ice are caused by the action of wind on snow. Such are the wave-like dunes called *sastrugi*. The long axes of these waves tend to lie in the direction of the prevailing wind and their size thus bears a relationship to the wind speed. Sastrugi can reach

Scientist exploring a crevasse near the northern edge of the Ross Ice-Shelf

Sastrugi

Frozen surface of a pond on an Antarctic glacier caused by weathering of the surface by wind and sun

Wind erosion of the surrounding snow has caused the snow compacted by these footprints to stand up in sharp relief

heights of up to 6 ft. and frequently form very solid obstacles; like crevasses they can be a formidable barrier to land transport, though they are far less dangerous.

Determining the thickness of the ice-sheet

Determining the actual thickness of the Antarctic ice-sheet is of primary importance. It enables the glaciologist to calculate not only the total volume of the ice-sheet as it is now but also to estimate the volume of ice that might have existed in the vast ice-sheets of the Northern Hemisphere which were very extensive some hundreds of thousands of years ago. Profiles, or cross-sections, of ice thickness can be used to compile diagrams showing the configuration of the underlying rock surface and with some interpolation of data an outline of the land mass beneath the ice can be drawn. The physical obstacles are formidable; scientists from the United States' Cold Regions Research and Engineering Laboratory have succeeded in drilling through 7100 ft. of ice to the rock at Byrd Station. Long before man began to explore the Antarctic ice-sheet scientists were using indirect methods to guess at its thickness, based on the fragmentary information brought back by coastal expeditions; such estimates ranged from 1400 ft. to 24 miles. Today's estimated average thickness is between 6000 and 8000 ft. Fortunately it is no longer necessary to guess, however intelligently, for science and technology have combined to provide a number of methods for probing the thickness of ice-sheets.

The first of these, the seismic method, is by far the most accurate as well as the most informative. It is a technique borrowed from oil-prospectors who use it to explore the nature of underlying layers of rock. Basically it consists of firing a charge of dynamite on the surface of the ice and picking up the sound impulses reflected from the bedrock by means of geophones, a type of microphone. The impulses from these are then amplified and passed to galvanometers whose movements are automatically recorded by a camera. These recordings provide a measurement of the time taken by an impulse to pass through the ice layer and return to the surface. From the travel time it is possible to make a direct calculation of the depths of the reflecting surfaces. This technique has been extremely successful in measuring ice thickness from which the outline of the underlying rock surface can be traced. The same equipment can also be used to give information from much greater depths (up to 30,000 ft.) on the physical nature and thickness of the various layers that make up the Earth's crust. Originally used by glaciologists on Swiss glaciers, the seismic technique was first pioneered in the Antarctic during Admiral Byrd's Second Antarctic Expedition of 1933–5, and then in Dronning Maud Land by the Norwegian-British-Swedish Expedition of 1949–52. But it was not until the all-out scientific effort which began with the International Geophysical Year of 1957–8, that the logistical support needed to carry the scientists and their heavy equipment over many thousand miles of difficult country became available. Undoubtedly the most dramatic seismic traverse of this period was that carried out by Sir Vivian Fuchs's team on the Commonwealth Trans-Antarctic Expedition in 1957–8; depth measurements were made at 44 separate points along the traverse. This was all the more remarkable when every hour was precious for seismic shooting is a laborious and time-consuming procedure. Up to five hours may be needed to set up and dismantle the apparatus and this allows only occasional shots to be fired on a long journey.

A simpler method of measuring ice-sheet thickness involves measuring the force of gravity which varies on the Earth's surface according to the distance from the centre. Gravity can be measured in several ways. The pendulum is the most accurate instrument though the least convenient to use in the field. Since the time taken by a pendulum to swing backwards and forwards varies with the strength of gravity, by timing thousands of swings at different places the difference in period, and thus the difference in gravitational force at different points, can be calculated accurately.

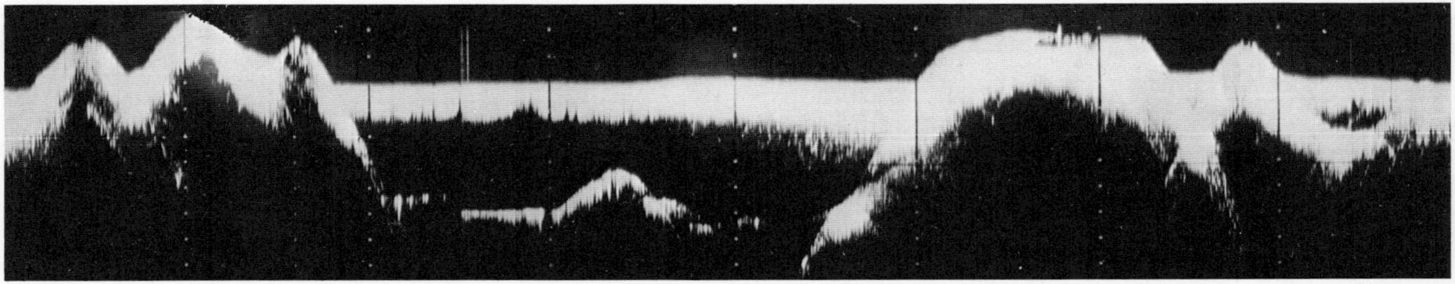

Cross-section of the Nimrod Glacier recorded by Scott Polar Research Institute radar transmitter-receiver. At this point the glacier is 2800 ft. thick and 11 miles wide. The rock ridge revealed in the centre is about 820 ft high

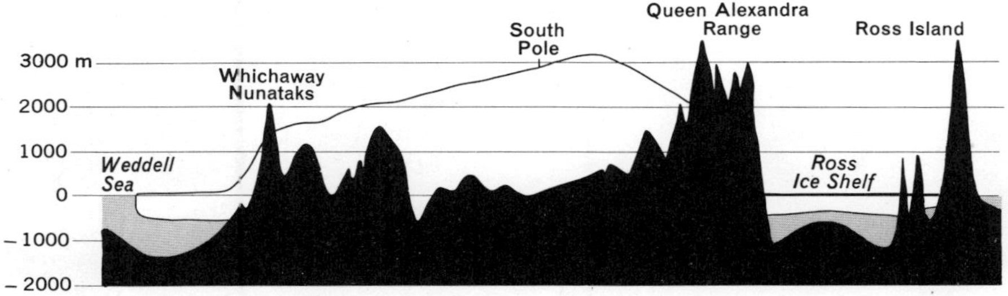

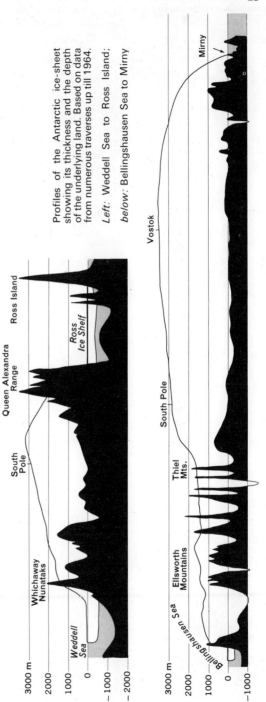

Profiles of the Antarctic ice-sheet showing its thickness and the depth of the underlying land. Based on data from numerous traverses up till 1964.

Left: Weddell Sea to Ross Island;

below: Bellingshausen Sea to Mirny

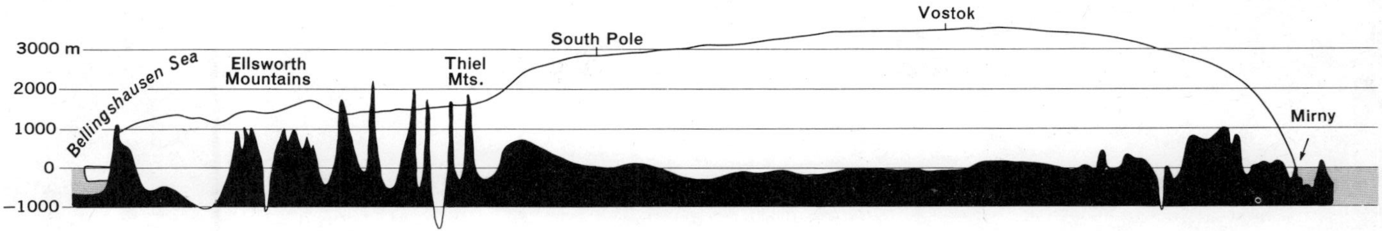

Left: Depth sounding of the ice-sheet by seismic shots

Right: British Antarctic Survey field party on a depth-measurement traverse of the ice-sheet using a radar transmitter-receiver mounted in a Muskeg tractor. The snow cairn is used for navigational purposes

Less cumbersome is the gravimeter, virtually a very accurate spring balance consisting of a small weight suspended from a thin spring of quartz and weighing only a few pounds. With this the geophysicist calculates from his results not the total force at any point but the gravitational anomaly, that is the discrepancy between the actual reading and the theoretical value at that point. Gravity measurements are a useful method of supplementing and checking depth figures obtained from seismic reflections and for tracing the profile of the rock surface between seismic stations.

Left: Close-up of radar transmitter-receiver developed by the Scott Polar Research Institute

Right: Weighing snow samples from a coring auger. The glaciologist to the right records snow density data

A technique for the continuous measurement of the thickness of the Antarctic ice-sheet has been developed at the Scott Polar Research Institute in Cambridge and promises to speed up the present rather slow and costly methods. This is a small radar transmitter–receiver mounted on a tractor or an aircraft. Instead of sound waves, radio-frequency waves are transmitted through the ice and a measurement of the velocity of the reflected waves is used to measure ice depth. The apparatus automatically records the depths of the ice on a moving photographic film. Successful measurements have already been made on the Greenland ice-sheet and in Antarctica. The advantages of a continuous measurement as distinct from the spot measurements provided by seismic and gravitational techniques are obvious. A bold imagination can even envisage such a radio echo-sounder mounted on a satellite to make continuous circuits of the Antarctic continent.

We have considered at some length the various techniques that glaciologists use to sound the depth of the Antarctic ice-sheet. Since the beginning of the International Geophysical Year of 1957–8 over 50 traverses have been made employing these techniques. Now the combined data, together with the available figures for surface elevation, make it possible to reconstruct rough maps of the underlying land mass, to calculate the average thickness of the ice-sheet and from this estimate the volume of ice. The average thickness of the ice-sheet, as we said at the beginning of this section, is between 6000 and 8000 ft. If we add to this the ice-shelves a total volume of ice is obtained equal to 2 per cent of the world's water budget and 90 per cent of all the ice in the world. Were this to melt sea-levels would rise by 200 ft. inundating our ports and low-lying regions. This is an awe-inspiring prospect and one which leads us to our next consideration – is Antarctica's ice-sheet growing or shrinking?

The Antarctic ice budget

One of the chief problems taxing the ingenuity of the glaciologist is that of determining whether the ice-sheet of Antarctica is getting thicker, wasting away or in a state of balance. If accurate data were available relating to the net accumulation and the net wastage of snow and ice then it would be possible to calculate the mass budget. But the technical difficulties of obtaining this information for such a vast and inaccessible region are formidable and, as things are, the most we can expect are a few well-informed guesses. The first problem is that of accumulation and how to measure it. Precipitation in Antarctica is almost entirely in the form of snow and, except in a few coastal regions, rain is virtually negligible. The distribution of snow is not evenly spread over the continent; the bulk of the snowfall is round the coast and it diminishes steadily towards the centre with almost desert conditions over the high inland ice-sheet. Snow is carried considerable distances by the wind and so it is well-nigh impossible to measure the amount falling with snow-gauges. In fact, falling snow can only be estimated by measuring the annual accumulation.

There are a number of ways of measuring accumulation; the simplest is to set up a network of stakes against which changes in the level of snow surface can be measured; this is the procedure normally used at established scientific stations. Where this is impracticable, on journeys across the ice-sheet for example, the usual practice is to dig a pit in the snow surface several feet deep and compare the annual layers of snow which can be read off like a scale. Since snow laid down in the summer months tends to be softer and coarser grained than the winter layers which are harder and more compact, it is possible to count the annual layers over a considerable period of time. At the South Pole layers to a depth of 88½ ft. and 200 years old have been measured by this technique, known as 'snow stratigraphy'. By using specially designed drills ice-cores can be obtained from much greater depths; one such frozen archive from a bore-hole 1014 ft. deep was estimated to be some 1700 years old, representing snow laid down in the heyday of the Roman Empire. On some parts of the ice-sheet, where snow stratigraphy is unreliable, there remain other and more intricate methods, such as comparing the ratio of the isotopes of oxygen in snow, which vary with the temperature, or measuring the age of the carbon dioxide in the air trapped in the ice. But when all existing techniques for measuring net snow accumulation have been brought into play and tested against numerous theories the answer is still subject to a 25 per cent error. One expert estimate gives a figure of $1.8+0.9$ million million tons of water equivalent as the total net annual accumulation of snow over the whole of the Antarctic ice-sheet.

The second problem to be resolved in computing the ice-sheet budget is wastage. There are four ways in which an ice-sheet can lose ice:

Firstly, wastage through calving of icebergs from the edge of the ice-sheet; in order to estimate the amount of ice lost each year by this means two facts are needed – the thickness of the ice and the rate of flow. During the past few years measurements of ice thickness and flow have been made in coastal regions of the inland ice-sheet, on glaciers, on ice-streams and on ice-shelves and it appears that ice-shelves account for the bulk of iceberg calving followed by ice-streams and lastly by the relatively slow-moving ice-sheet.

Secondly, wastage caused by *ablation*, the term given to loss by evaporation and by run-off of meltwater. In some windy areas the conversion of snow into water vapour is a significant occurrence but elsewhere this process tends to be balanced by condensation of moisture from the atmosphere on to the snow surface.

Thirdly, wastage due to blown snow; many million of tons of snow are blown out to sea every year.

Fourthly, bottom melting; temperature measurements made in holes drilled in floating ice-shelves suggest that there is a certain amount of melting occurring at the bottom towards the seaward boundaries of ice-shelves.

The total estimates of ice loss seem to vary even more widely than those for ice gain. One calculation will show a net loss, another a balanced budget. There are various indirect methods of checking the answer to this problem. One would be to measure the level of the ice-sheet relative to protruding nunataks, or mountain-peaks, but these are relatively scarce, especially on the inland ice. Another method might be to check accumulation at particular spots by means of seismic depth sounding; unfortunately the technique is not sufficiently accurate to detect the small annual changes caused by the addition of a few inches of snow. A third method might be to observe changes in world sea-levels. Sea-levels do in fact appear to be rising which might suggest that Antarctica is on the melt. Unfortunately, thermal expansion can also account for rising sea-levels, so that this method of comparison would also seem to be inconclusive. All that can be said for certain at this stage is that the ice-sheet of Antarctica seems to be stable; meanwhile, we must await the results of further and more detailed budget studies.

3
THE LAND

Ice, as we can now appreciate, is the predominant feature of the landscape of Antarctica; it covers perhaps 95 per cent of the underlying rock. The remaining 5 per cent has a double scarcity value for the scientist; firstly, because he needs it to use as a solid foundation upon which to build living quarters and laboratories; secondly, because the ice-free slopes of mountain ranges, the exposed tips of buried peaks, or *nunataks*, and the rare iceless valleys form as it were windows into the structure of the continent itself. Among the items of equipment found by the dead bodies of Captain Scott, Edward Wilson and 'Birdie' Bowers after their disastrous return journey from the Pole was a sledge on which were 35 lbs. of rock specimens, a silent testimony to the importance they attached to what to many might appear to be uninteresting lumps of stone. For at this time in the history of Antarctic exploration little was known about the geological structure of the continent. No one could be sure whether it was one land mass or maybe two or even a number of scattered islands welded together by one ice-sheet. Since that era of pioneers conjecture has given place to certainty. Our purpose in this chapter is to review the present state of knowledge of the shape of sub-ice Antarctica, the history of its stratigraphy or succession of rock formations and its relationship to the neighbouring continents. And finally we must attempt to answer the question put by those who seek to find an economic justification in Antarctic exploration: 'Has Antarctica any minerals of economic importance to the rest of the world?' This is the general picture we want to outline, but before doing so it would be useful to know just where this small 5 per cent of ice-free rock, on which the expert must base his generalizations, is to be found.

Mountains

There are two main regions of Antarctica where the landscape is dominated by rock rather than by ice; they are Victoria Land in Greater Antarctica and the region of the Antarctic Peninsula in Lesser Antarctica. In Victoria

Nunataks near Borga, Dronning Maud Land

Land the Royal Society Range dominated by Mount Lister (13,353 ft.) provides a magnificent backcloth to the whole historic region of McMurdo Sound. Flanked by the Admiralty Range in the region of Cape Adare and the Queen Alexandra and Queen Maud Ranges stretching towards the Pole, they form but a section of one of the world's greatest mountain chains, the Transantarctic Mountains stretching from Oates Land to the Filchner Ice-Shelf and onwards in widely spaced groups through the Shackleton Range to the Theron Mountains in Coats Land and possibly even into Dronning Maud Land, a total distance of 3000 miles. Other major mountain systems ringing Greater Antarctica are to be found in Dronning Maud Land from the Weddell Sea to Lützow-Holmbukta and including the Mühlig-Hofman and Wohlthat mountains, the Sør-Rondane mountains, the Belgica mountains and the Queen Fabiola mountains. In adjacent Australian Antarctic Territory are the numerous high peaks of Enderby Land which include the Tula Range with its main peak Mount Riiser-Larsen rising to a height of 6107 ft. Flanking the Lambert Glacier in Mac.Robertson Land are a number of ranges of which the Prince Charles range (named after the heir-apparent to the British throne) is the major group. Along the coast of Princess Elizabeth Land is a 200-square-mile region of mainland rock and off-shore islands known as the Vestfold Hills. Other isolated mountain outcrops occur round the coast into Oates Land.

The Antarctic Peninsula, in particular the west coast of Graham Land and its off-shore islands, provides some of the most magnificent scenery in the whole region, its fjords and mountains being reminiscent of the coasts of northern Norway. South of the peninsula, in Byrd Land, lie the Ellsworth Mountains named after Lincoln Ellsworth the United States

Oblique air photograph of Sentinel Range showing Vinson Massif, highest mountain in Antarctica, and Mount Tyree

Mount Tyree in the Sentinel Range of the Ellsworth Mountains, second highest peak in Antarctica

aviator who discovered their northern sector, the Sentinel Range, during his transcontinental flight in 1935. The highest peaks in this range are also the highest peaks in Antarctica, Mount Tyree (16,290 ft.) and the Vinson Massif (16,860 ft.). The southern sector of the Ellsworth Mountains, the Heritage Range, is lower and less spectacular. Some geologists believe that the Ellsworth Mountains belong to an intermediate province between the younger rocks of Lesser Antarctica and the older formations of Greater Antarctica. Extending along the coast of Byrd Land towards the Ross Ice-Shelf are a number of isolated mountain ranges. Such are the Jones Mountains on the Eights Coast of the Bellingshausen Sea, a mere series of nunataks whose volcanic origin links them with the mountains of the Antarctic Peninsula. The Crary Mountains, the Executive Committee Range and the Edsel Ford Range appear to have similar geological affinities which suggest a possible common origin.

Volcanoes

A prominent landmark in the McMurdo Sound area is Antarctica's only recently active volcano, Mount Erebus (3743 ft.). Its changing colours and ascending plume of smoke inspired the brush of Edward Wilson, doctor and scientist on both Scott's *Discovery* and *Terra Nova* expeditions. Erebus is one of several volcanoes which have coalesced to form Ross Island. Other volcanic islands in the vicinity include Scott Island and the Balleny Islands. The drowned volcano called Deception Island, in the South Shetland group, is a reminder of the volcanic origin of this part of Antarctica.

Ross Island, McMurdo Sound, formed by three volcanoes, Mounts Bird, Erebus and Terror (*left to right*), with Hut Peninsula in the foreground

Crater of Mount Erebus

Masson Range, near Mawson Station, Mac.Robertson Land

Mount Erebus from Rocky Point, near Cape Royds, Ross Island

Peaks of the Yamato Range, Dronning Maud Land

Typical scenery of western Graham Land, Antarctic Peninsula

The mountainous west coast of the Antarctic Peninsula with Neny Island in the foreground lying at the mouth of frozen Neny Fjord

Oases

During the summer months there are a few places round the coast of Antarctica which are partly or wholly ice free; such areas are valuable not only to geologists but to naturalists as well as they are often frequented by many kinds of plant, insect and animal life. Such is Cape Hallett in north Victoria Land, Ross Island with its vast scientific station at McMurdo and the site of the Australian Mawson Station in Mac.Robertson Land. But by far the largest continuous areas of ice-free land in Antarctica are the oases, the most extensive of which is the McMurdo Oasis in Victoria Land. Lying in the shelter of the Royal Society Range, it is between 9 and 15½ miles wide and 93 miles long. It consists of three main valleys, once occupied by glaciers, the Taylor Dry Valley, the Wright Valley and the Victoria Valley, as well as a number of smaller valleys. To the south-east and the north of the main oasis lie lesser ice-free regions. The term 'oasis' is not really very appropriate in this context; far from being havens of fertility in a desert of ice, these valleys have exceedingly desert-like climates; precipitation is low and summer temperatures close to freezing. Many reasons have been put forward to explain the presence of the dry valleys, the most probable being that the katabatic, or downward-falling winds from the Antarctic plateau, compress and therefore raise the temperature of the bottom air of the valleys; coupled with this is the fact that large areas of exposed rock will absorb more solar radiation and thus help to raise the average temperature. Clearly the valleys are warmer because they are ice free rather than the other way about. Because temperatures are low and precipitation scarce there is no great abundance of plant life other than algae, lichens and mosses. None of the seals, penguins and sea birds so prolific along the adjacent coastal zone are to be found here; but the presence of mummified seals as far inland as 37 miles from the sea at an elevation of 3000 ft. has puzzled zoologists for some time. The simplest explanation of these curious erratics in the dry valleys is that at some time in the past they must have wandered inland from their native beaches and then somehow lost their sense of direction and died of starvation. The same fate has also overtaken a number of Adélie Penguins whose corpses have been found in the valleys.

Unique features of these unique valleys are the saline lakes, Lake Vanda in the Wright Valley and Lake Bonney in the Taylor Dry Valley; these are fed by streams originating from melting glacier snouts. Both are ice covered yet temperatures at the bottom are considerably higher than temperatures at the surface; the deepest water in Lake Vanda (249 ft.) has a maximum temperature of 26 °C. and the gradual increase in temperature from surface to bottom appears to correlate with a gradual increase in salinity; whereas you can drink the water in the top 170 ft., the bottom water is extremely salty. The source of the heat is almost certainly the Sun, the rays of which are focused on the bed of the lake by surface ice

Rocks hollowed by the action of the wind

crystals. The cause of the high salt concentration may be due to a former less saline lake which evaporated and left a high concentration of salts behind.

Ice-free valleys, like the Wright Valley, are of particular interest as places in which to study various post-glacial processes such as erosion, the wearing down of exposed rocks by the elements or by chemical action. The curiously shaped lumps of rock, reminiscent of *avant-garde* sculpture, known as *ventifacts*, are scattered over the floor of the Wright Valley. On one side their hard, shiny polished surfaces have been sand-blasted by the prevailing winds; on the other side they will be soft and crumbly, weathered into pits and honeycombs.

There are other and lesser oases elsewhere in Antarctica such as the Bunger Hills in Wilkes Land and the Vestfold Hills in Princess Elizabeth Land. Oases are also to be found in the Antarctic Peninsula and its off-lying islands but these are only a few square miles in area.

The land beneath the ice

The fact that 95 per cent of the land of Antarctica is obscured by ice has not prevented scientists from reconstructing the general pattern of the sub-ice continent and outlining its geological history. Where the rocks come to the surface no special problem is posed other than the logistic one of transporting and supplying the geologists in the field. Where the rock

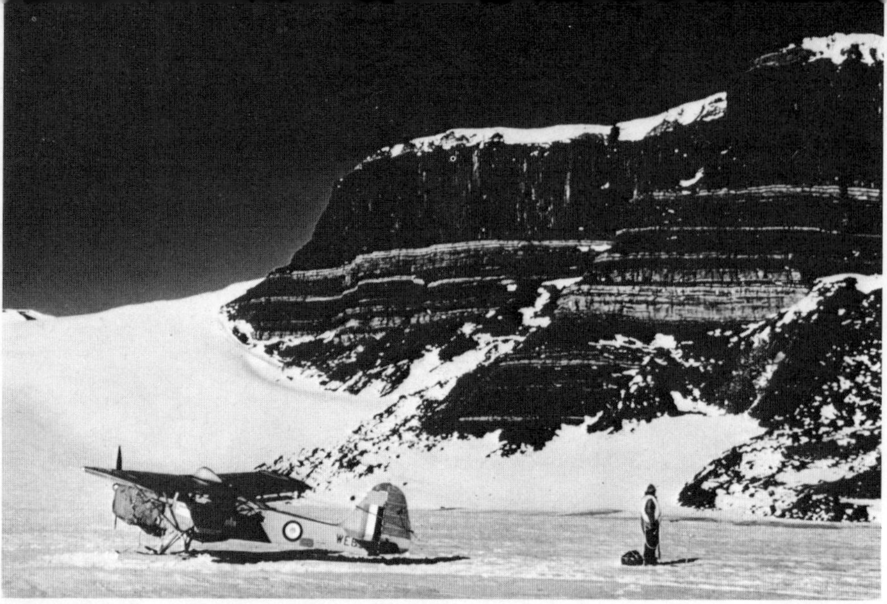

Theron Mountains containing coal seams assigned to the Permian period and corresponding to the main coal horizons of Australia and South Africa

face is under thousands of feet of ice then the problem of exploration is a formidable one. The only direct approach is to drill down through the ice to the bedrock and a drilling-rig has now been designed in the U.S.A. which can penetrate to over 7000 ft. But drilling is a slow process and in order to obtain sufficient data over many thousands of miles indirect techniques must be used; these are in practice identical with those already described for measuring the depth of the ice-sheet including seismic surveys and radar surveys. From the data obtained by these techniques it is possible to calculate the height of the land beneath the ice in relation to the surface of the ice-sheet and to sea-level and also to learn something of the nature of the bedrock itself. Information about the hard rock crust upon which the continent floats in the Earth's yielding mantle can be gained from the study of earthquake waves originating from other parts of the world. There are now a number of seismic stations in Antarctica recording earthquake activity, though the continent itself is what is termed 'aseismic', that is no earthquakes of any appreciable size have ever been recorded there; this is curious for during its geological history the region has experienced several periods of volcanic activity, normally an indication of crustal instability. A possible explanation may be that the weight of the ice-sheet by depressing the continent 600 ft. on average is in some way responsible for suppressing earthquake activity.

Supposing all the ice now covering Antarctica were to be removed, how would the sub-ice continent appear? Sufficient information has been gathered together in the last ten years or so to provide a rough answer. Greater Antarctica would appear as a low-lying land mass characterized by a

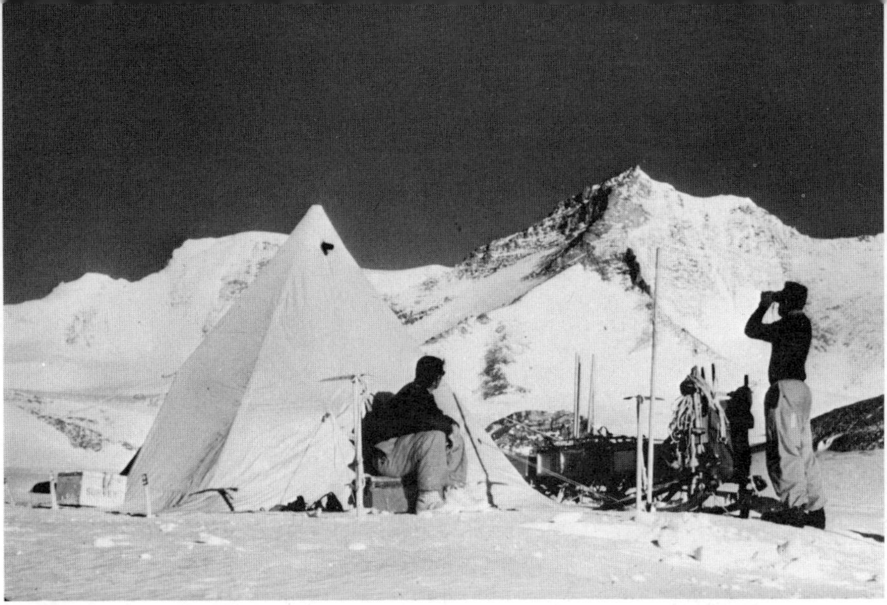

Geological field party in the Tottanfjella, a range of mountains in Dronning Maud Land

great central plain bounded on one side by the Transantarctic Mountains and stretching from Victoria Land to Queen Mary Land. Lesser Antarctica would be greatly shrunk in area, the rugged mountains of the peninsula remaining but Byrd Land being reduced to a number of archipelagos. Even allowing for isostatic uplift, that is the amount the continent would rise in the Earth's mantle if the overburden of ice were to be removed, the average height of Antarctica would be much less than it is now, perhaps only 3000 ft. For many years geologists postulated a great sub-ice trough linking the Ross and Weddell Seas and dividing Greater from Lesser Antarctica; this is now shown not to exist. Instead a channel 300 to 350 miles wide and reaching a depth of some 8000 ft. below sea-level joins the Ross and Bellingshausen Seas with a branch possibly linking up with the Weddell Sea.

Without its ice-sheet Antarctica seems to be structurally no different from any other continent. An interesting contrast, however, lies in the nature of its continental shelf, that gradually sloping continuation of a land mass into the sea which is the true geological boundary of a continent. The depth of Antarctica's continental shelf has for long been known to be greater than that of other continents – anything from 1600 to 2600 ft. below sea-level compared to 650 ft. elsewhere. The cause of this difference has often been attributed to the weight of the ice pressing the continent down into the Earth's mantle, but gravity measurements seem to show that while the ice-laden central parts of Antarctica are indeed depressed, the coastal regions which happen to be free of ice – and by extension their shelves – are not depressed in this way. So the problem remains unsolved.

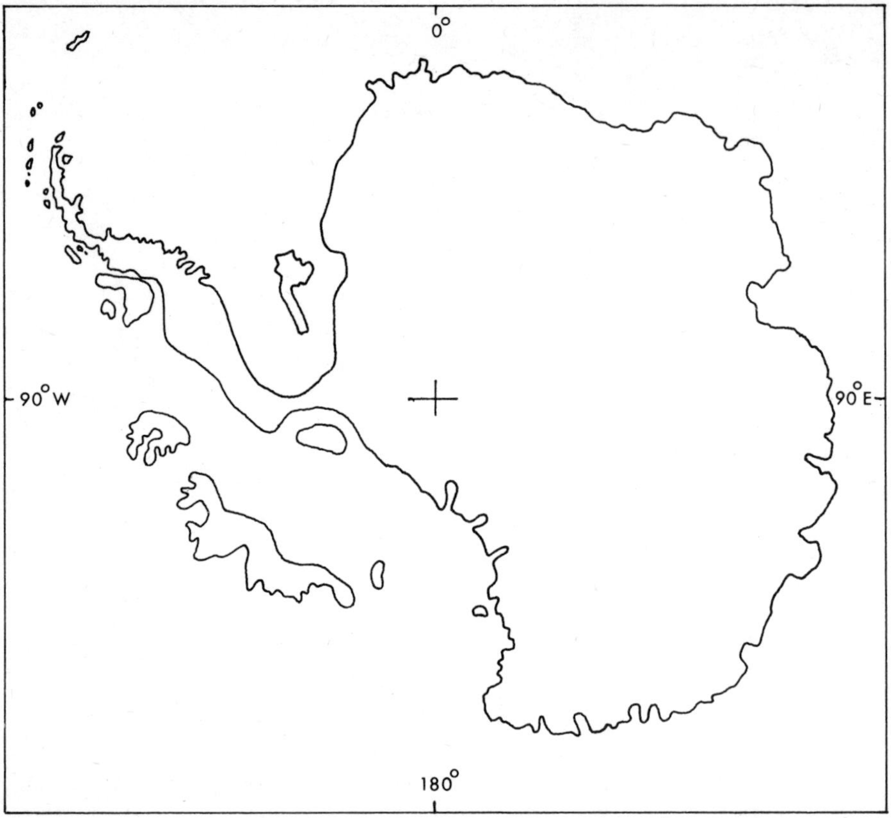

The approximate outline of Antarctica after removal of the ice-sheet and subsequent isostatic compensation. Much of Lesser Antarctica is below sea-level, while Greater Antarctica is entirely above sea-level

Geological history

Not only have we now a much clearer picture of the shape of the Antarctic land mass but its geological history and structure are likewise becoming clearer. There was a time when it was difficult to do more than assign relative ages to such of Antarctica's rocks as were accessible to the geologist's hammer, but nowadays many rocks can be dated by advanced physical and chemical techniques which involve measuring the amount of radioactivity in the crystals that form them. The principle of radioactive dating of rocks is based on the fact that the isotopes* of certain elements – uranium and thorium for example – are in a constant state of atomic disintegration. The disintegration of the atoms in time alters the nature of the original element. The process can continue through a number of successive radioactive elements until finally an end-product is reached which is non-radioactive. In the case of uranium these end-products are helium and lead.

* Isotopes are atoms of the same element having the same atomic number but differing in atomic weight.

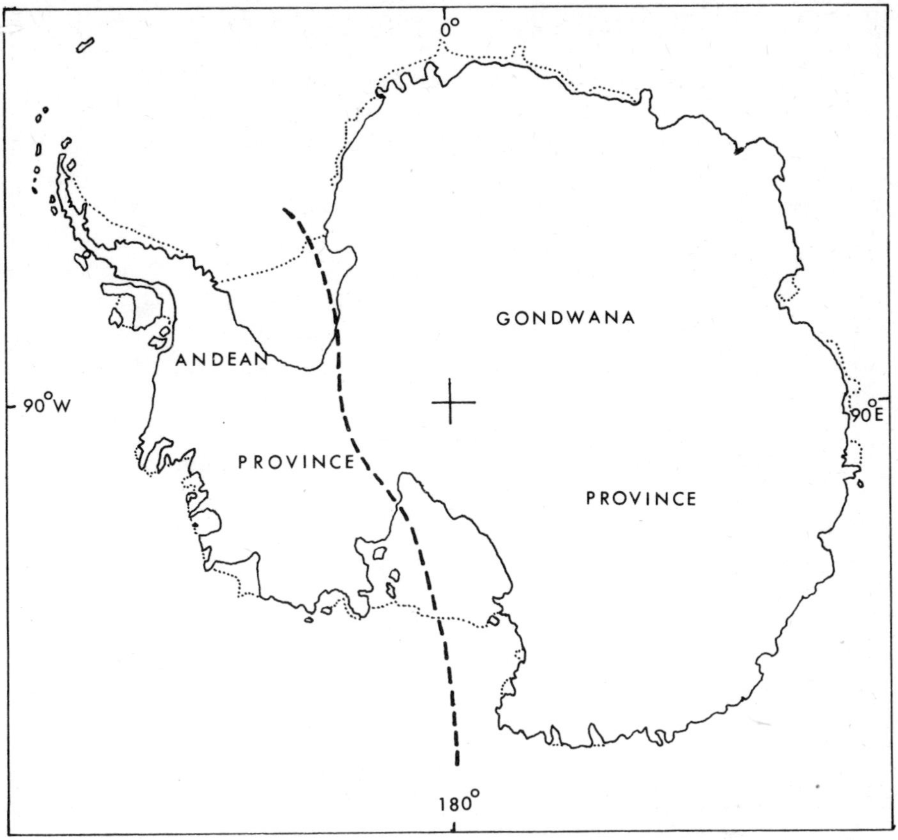

The approximate boundary between the Andean and Gondwana geological provinces of Antarctica. A comparison with the figure opposite shows the relationship between surface topography, sub-ice topography and geological structure

The rate at which uranium decays is known and by estimating the amount of uranium-lead in a mineral it is possible to calculate the age of the parent rock. A similar method depends on argon generated by a radioactive form of potassium. The oldest absolute age yet determined by this technique is in the region of 1600–1500 million years in the Vestfold Hills area of Princess Elizabeth Land and in Terre Adélie, regions 600 miles apart.

The stratigraphy of Antarctica, that is the descriptive study of its various rock strata, is complicated by the fact that geologically speaking Antarctica is not one continent but two. The larger of these two continents corresponds to Greater Antarctica whose ancient basement of rocks crystallized, as we have seen, in one of the older periods of geological time. This Gondwana Province as it has been called after the conjectural continent of Gondwanaland of which Antarctica, it is thought, once formed a part, is structurally different from the Andean Province which is welded on to it. Roughly 400,000 square miles in area, this latter region, whose much-folded structures are similar to and indeed a continuation of the folded structures of the Andes of South America, includes the Antarctic Peninsula and forms Lesser Antarctica; the exact boundary between the Andean and Gondwana Provinces has yet to be fully determined.

Greater Antarctica's oldest rocks date from the earliest period of the geological time-scale known as the Precambrian, that is approximately 4500–600 million years ago. The proportion of these basement rocks not buried by ice or rocks of a subsequent period is small and outcrops of them are relatively few and isolated but their distribution over the continent is widespread. The only evidence for life during this period are the tiny spore-like fossils found in Late Precambrian rocks all over the world. The long process of deposition of marine sediments continued for a further 100 million years or so into the Cambrian period (600–500 million years ago). Among rocks brought back by Sir Ernest Shackleton's *Nimrod* Expedition of 1907–9 from the Beardmore Glacier was a limestone boulder containing the oldest recognizable fossil to be found in Antarctica, an archaeocyathid, a coral-like organism also found in Cambrian rocks from other continents. These organisms suggest that warm-water conditions were characteristic of Antarctica at this time. During the Ordovician period, 500–440 million years ago, granite rocks pushed their way through the Cambrian sediments; such are the pinkish-red granites of the McMurdo Sound area; their formation and subsequent erosion marks the completion of the sequence of processes by which the basement rocks of Greater Antarctica were formed.

The younger rocks of the Gondwana Province were formed between 400–350 million years ago during the Devonian period; they were first investigated by Frank Debenham, a young geologist with R. F. Scott on the *Terra Nova* Expedition of 1910–13. Like the Carboniferous period which followed it, the Devonian was a period of deposition; a textbook example was investigated some years ago by a team of American geologists in the Ohio Range of the Horlick Mountains, Byrd Land. Here they found Devonian sandstones containing fossils of primitive whelks and snails as well as plants, suggesting that the climate during this period was relatively mild. Then came evidence of Antarctica's first recorded ice age some time between the end of the Carboniferous and the subsequent Permian period, that is some time within the bracket 300–250 million years ago. Overlaying the Devonian rocks the geologists found a 900 ft. thick layer of tillites or rubble consolidated into a mass by an advancing glacier. In some places the Devonian rock was swept away exposing the underlying basement rock. This ice age, which may well have been a major one, was not that which we see in Antarctica today, for above the tillites were Permian fossils which contained the petrified leaves of *Glossopteris*, a large-leafed fern, a plant characteristic of much of the Southern Hemisphere during the period following the glaciation. Associated with *Glossopteris* in the

Geologists in action, Anvers Island, Antarctic Peninsula

Ohio Range were coal-beds 13 ft. thick. For coal to have formed there must have been a warm climate with an abundant and luxuriant vegetation. Trees and plants must have fallen into swamps to form first peat and later coal; indeed petrified tree-trunks have been found in the sandstone measuring 24 ft. in length and 2 ft. in diameter. Similar evidence has been found from other parts of Greater Antarctica; during Scott's *Discovery* Expedition of 1901–4 his geologist, H. T. Ferrar, found *Glossopteris* fossils and coal-seams in the sandstones of Victoria Land which he called the Beacon Sandstones. Until very recently no land vertebrate fossils had been found in Antarctica though the fossilized imprints of their tracks had been reported. Then in December 1967 a team of geologists from the Ohio State University's Institute of Polar Studies found the missing link – part of a bone of a fossil land vertebrate. This has been identified as a piece of the jawbone of an amphibian of the now extinct subclass Labyrinthodontia and was found in rocks of the Triassic period, about 200 million years ago.

Following this warm or even semi-tropical period the climate seems to have grown arid and desert-like during the Triassic period to be followed during the Jurassic period 180–135 million years ago by an age of catastrophic volcanic activity. The geological record of Greater Antarctica then comes to an abrupt halt; the sediments associated with the succeeding 135 million years, that is the periods of the Cretaceous, 135–70 million years ago and the Tertiary, 60–2 million years ago, are largely hidden by the continental ice-sheet. Only fragmentary evidence has been found for the Tertiary period, fossils of microplankton, algae and plant pollen from the McMurdo Sound region; the plant pollen, on analysis, suggests a cool temperate climate for this period. Some evidence which helps to fill this big gap is to be found among the more exposed rocks of the Antarctic

AGE IN MILLIONS OF YEARS	PERIOD	ERA	
1 million years	QUATERNARY	CENOZOIC	Formation of present ice-sheet.
50	TERTIARY	CENOZOIC	Antarctica roughly its present shape. Volcanic activity in Andean Province. Mountains of Antarctic Peninsula raised. Fossils suggest temperate climate.
100	CRETACEOUS	MESOZOIC	Period of deposition and mountain-building in Andean Province.
150	JURASSIC	MESOZOIC	Period of volcanic activity in Gondwana Province. Fossils from Andean Province suggest luxuriant plant-life.
200	TRIASSIC	PALAEOZOIC	Desert-like conditions probably prevailing. Period of deposition with coal deposits. Fossil trees and plants suggest warm climate.
250	PERMIAN	PALAEOZOIC	Period of glaciation.
300	CARBONIFEROUS	PALAEOZOIC	Further erosion and deposition.
350, 400	DEVONIAN	PALAEOZOIC	Period of deposition of sandstone and shale. Younger rocks of Gondwana Province formed. Fossil plants and shellfish suggest mild climate.
	SILURIAN	PALAEOZOIC	Further deposition and subsidence, folding and mountain-building.
450	ORDOVICIAN	PALAEOZOIC	Granite rocks intrude Cambrian sediments.
500, 550	CAMBRIAN	PALAEOZOIC	Process of deposition of sand and thick mud on sea-floor continues. Archaeocyathid fossils from limestone of Victoria Land suggest warm-water conditions.
600 – 4,500	PRE-CAMBRIAN		Period of much mountain-building and deposition.

GEOLOGICAL TIME-SCALE OF ANTARCTICA

Peninsula and its adjacent islands which form the basis of the later Andean Province.

The geology of the peninsula differs itself in many ways from that of the continent proper. Basement rocks, formed in Precambrian or Cambrian times, are only to be found in the southern half of the peninsula south of Marguerite Bay. The Cambrian and Jurassic periods are marked by some pink and white granites from the region of Marguerite Bay but the Ordovician and Silurian periods do not appear to be represented at all. A group of sediments, 40,000 ft. in thickness in Trinity Peninsula, the northern tip of this 'panhandle', and known as the Trinity Peninsula Series, may belong to the Carboniferous period, but unfortunately they are very poor in fossils. The Jurassic, which as we have seen in Greater Antarctica was a period of luxuriant plant growth, is splendidly represented in the peninsula by the Middle Jurassic fossils found at Mount Flora, Hope Bay, on the northern tip of Trinity Peninsula. These fossils, lying in beds 10 to 100 ft. in thickness, include ferns and similar plants as well as snail-like creatures, aquatic beetles and the remains of fish. James Ross Island, off the north-east coast of the peninsula, also has rich fossil-bearing strata dated as Upper Cretaceous; here the fossils include snails, bivalves and even lobsters. It is evident from the way the fossil-bearing strata of the peninsula region are folded and overfolded by mountain-forming activity that the whole region has had a much more unstable later history than Greater Antarctica whose youngest strata tend to lie in flat horizontal planes. During the Tertiary period the same processes that raised the

Photograph and line drawing of fossil leaf *Glossopteris* from Terrace Ridge, Ohio Range. *(Photo by O. G. Oftedahl; line drawing by Gillian Bull and J. F. Rigby, U.S. Geological Survey, Columbus, Ohio)*

Andes of South America also built the present-day mountains of the Antarctic Peninsula. Though Tertiary rocks are not widely distributed over the area there are some very interesting rocks of this period on Seymour Island, at the north-east corner of the peninsula, where the remains of fossil penguins, far taller than those living there today, have been found. During the Tertiary period the climate must have been a warm one; fossil flora found at Hope Bay include the early forerunners of the conifer and beech, typical of the Southern Hemisphere at this time. But though the fossil evidence shows many plant and animal similarities between the southern continent and its neighbours it is from about the middle of Tertiary times that Antarctica's true period of isolation seems to date for the mammals which elsewhere began to predominate among the animals some 30 million years ago are quite absent here. Towards the end of the Tertiary, only a few million years ago, there occurred a period of considerable volcanic activity of which the volcanoes of the South Sandwich Islands and Mount Erebus on Ross Island are the only active survivors. By now the Antarctic continent must have achieved something like its present shape. At the beginning of our present period, the Quaternary, a million years ago, the climate again grew colder and an ice-sheet, very likely the one we see today, began to creep over the Antarctic landscape.

Antarctica and the theory of continental drift

From Antarctica's meagre 5 per cent of exposed rock geologists are gradually reconstructing the time sequence, the relationship and the fossil content of the various strata and are now able to describe in broad terms the general structural arrangement, that is the distribution of the mountain ranges and the intervening basins. We have already seen how geological history can be approximately dated by radioactive techniques. A great deal can also be learned by comparing the rocks and fossils of Antarctica with those of neighbouring continents. In the course of this comparison geologists have become aware of the remarkable resemblances between certain sedimentary deposits in South America, Africa, India, Australia and New Zealand. The fact that these continents can also be fitted together in a rough jigsaw – the east coast of South America and the opposite coast of Africa are a particularly good fit – has been a source of numerous theories for many years which suggest that at some time in the remote past what are now separate continents were a single unit which in some way broke up and drifted apart. As long ago as 1912 the German meteorologist, Alfred Wegener, suggested that the present continents were derived from one supercontinent Pangaea; this theory was later modified when two such continents were conjectured, a northern one, Laurasia, and a southern one, Gondwanaland (named after Gondwana in India). The theory of continental drift was at first regarded as eccentric and improbable; one of the chief objections was the lack of any convincing theory to explain the mechanism by which whole continents could be moved through the Earth's crust. The great strides made in our knowledge of the structure of the Earth's interior and mantle have now provided convincing explanations of these processes. It seems possible that the convection currents which are responsible for transferring the heat from the Earth's interior to the crust along the great submarine ridges, such as the Mid-Atlantic Ridge, could also bring about the disruption and moving of whole continents.

The problem of reconstructing the supercontinent Gondwanaland was hindered for many years by the lack of information from one of the keys to the puzzle, Antarctica. Early in the present century a relationship between the rocks of the Antarctic Peninsula and the Andes and between the rocks of Victoria Land and South Africa was demonstrated. But only comparatively recently has there been enough evidence digested combined with adequate geological mapping to demonstrate that the theory of Gondwanaland is most probably true. Let us review very briefly some of the evidence that geologists have assembled to enable this fundamental assertion to be made. Beginning with the older geological province of Greater Antarctica it appears that the order of the rock strata is exactly the same as that of southern Africa. Again the Precambrian rocks called 'charnockites', first found in India, are typical of much of Greater Antarctica while similar rocks have been found in Australia and southern Africa. Furthermore, the ages of the Antarctic rocks seem approximately to correspond with those of the other southern continents; the coal-seams which as we noted earlier in this chapter belong to the Permian period may well have formed part of one vast coal-bearing sequence of which parts still remain in southern Africa and Australia. The tillite, evidence of Antarctica's first recorded glacial epoch, sometime during the Carboniferous and Permian periods, is also found in the other southern continents and the fact that the tillites in each stage of the glaciation have differing ages suggest that possibly this resulted from the slow drift of Gondwanaland from east to west across the south polar regions, each continent being overwhelmed in turn by the ice.

Just when Gondwanaland began to break up is difficult to say; it might well have occurred during Jurassic times when volcanic activity, caused by fracturing of the Earth's crust, was prevalent in all the southern continents. It is not likely that the subdivision occurred all at once but took place over a period of 50 million years or so ending sometime in the Early Cretaceous period.

The existence of certain land plants in the southern continents – *Glossopteris* for example, which we know thrived in Antarctica from Devonian to Jurassic times – further strengthens the evidence for the existence of Gondwanaland. During this same span of time it appears that the floras of Antarctica and of the other southern continents were identical. Incidentally, the plants provide evidence that Gondwanaland

Reconstructions of Gondwanaland during the late Carboniferous period (*above*) and the early Cretaceous (*below*). Previous positions of the land masses are shown in white and present positions in black. The first map shows the continent at a time well before the first rifts had occurred when it was moving from east to west across the south polar regions. By the early Cretaceous period the continents had started to drift apart.

was much warmer than Antarctica is today. The recent discovery of a fossil bone from the Triassic amphibian Labyrinthodontia, referred to earlier, adds additional support for this theory; the amphibian, now extinct, belongs to a major group of amphibians ranging from alligators to salamanders in size, which were dominant on the adjacent continents 350 to 200 million years ago. This group may well have included the ancestor of all land vertebrates.

Yet further evidence is available from the study of palaeomagnetism, a technique based on the magnetic properties of some rocks. It is known that when a rock cools from the molten state the direction of the Earth's magnetic field becomes fixed within it. By calculating the position of the magnetic pole at the time of this event the original latitude of the rock can be estimated. From this kind of evidence gathered over the past 15 years we can now approximate the gradual drift of the southern continents.

And finally there is the evidence of palaeoclimatology, or ancient climates. We have already seen how Greater Antarctica underwent a number of climatic fluctuations, how during the Carboniferous and Permian periods an ice age overwhelmed the land and how in the Jurassic a warming period developed accompanied by a luxuriant flora; and how this in turn was succeeded by desert conditions. These fluctuations seem to have been paralleled in the other southern continents; recent studies seem to suggest that as Gondwanaland drifted so each continent in turn was overtaken by a succession of varying climates. In this climatic sequence Antarctica has been placed provisionally between India and Australia.

In a world whose technological growth looks like outstripping the available supply of economic minerals it is natural to think of Antarctica as a possible reserve supply. What minerals has Antarctica to offer and what use can we make of them? Over 200 different minerals have been listed to date though many of these have been detected in only minute quantities. Within the older shield area of Greater Antarctica potentially valuable minerals include those containing antimony, chromium, copper, gold, lead, manganese, molybdenum, tin and zinc. Coal-bearing strata exist in Lesser Antarctica and it is possible that one day oil may be found in the region of the Antarctic Peninsula. So far no minerals have been found in any appreciable quantity. The economics of mining ore from the depths of the Antarctic ice-sheet and extracting the minerals on the site, added to the cost of shipping them to distant parts of the globe through the world's most perilous waters make it unlikely that any future discoveries will add significantly to total production though they may act as a reserve against a future scarcity.

4
THE CLIMATE AND WEATHER

One fact with which everybody is familiar about the climate of Antarctica is that it is cold, cold enough as we have seen in a previous chapter to maintain in a state of equilibrium an ice-sheet containing some 90 per cent of the world's total amount of snow and ice and cold enough to freeze the surface of the surrounding ocean for a distance of several hundred miles in winter. Antarctica's climate is certainly very much more rigorous than that of the Arctic and its influence affects a very much larger area; in this respect it is unusual. Nevertheless, meteorologists do not study Antarctica in isolation, for along with the Arctic it is an essential component in the great heat exchange mechanism which causes the world's weather. In this chapter we attempt an explanation, with the minimum of technicalities, of the working of this mechanism and the part played by Antarctica in its operation. Then we shall consider the main characteristics of the Antarctic climate and the types of weather conditions which face the explorer in the field.

Definitions

Many terms which meteorologists use very specifically when describing their subject are also used by the man-in-the-street in loose and often contradictory ways. Many of these words like 'climate' and 'weather', for example, appear in profusion throughout this book; we need to be quite clear, then, as to their meaning. Thus, 'meteorology', with which this chapter is wholly concerned, is derived from the Greek words *meteoros* (lofty, elevated) and *logos* (discourse) and means the science of the atmosphere. The atmosphere is the gaseous envelope surrounding the Earth, held to it by gravity and largely rotating with it. The atmosphere is divided into a number of layers which include the troposphere, the stratosphere and above these the ionosphere. The troposphere and the stratosphere we shall be discussing later on in this chapter. By 'weather' we mean atmospheric conditions as they affect human activities – temperature, precipitation (i.e. snow, rain, hail, sleet), cloudiness, wind and so on, considered over short-term periods and measured in minutes, weeks and months. 'Climate' on the other hand is the synthesis of all these elements for a particular locality over a much longer period of time, usually about 30 years.

History

The first meteorological observations made in the Southern Hemisphere south of latitude 50°S. date back to the eighteenth century. But observations from Antarctic land stations begin only with the present century and for the first 50 years or so were limited to the coastal regions of the Antarctic Peninsula and such accessible parts of the coastline as McMurdo Sound in the Ross Dependency. Not until the International Geophysical Year of 1957–8 was a permanent weather-station established on the inland ice-sheet. Today a fully co-ordinated network of weather-stations, both round the coast and inland, including one at the South Pole itself, gather meteorological data all the year round. This is passed to an International Antarctic Meteorological Research Centre in Melbourne, Australia, where atmospheric data are analysed and incorporated in daily synoptic charts covering the whole of the Southern Hemisphere. Climatic data in this form are valuable for such long-term climate projects as the planning of air routes in the Southern Hemisphere, the measurement of the heat and moisture exchanges involved in the world distribution of climates, and the study of past climates. There are still large gaps in the coverage of Antarctic weather-stations which need to be filled, especially in the Southern Ocean where so far it has not been possible to maintain regular weather-ships.

Why the poles are cold

Sufficient meteorological data is now available for experts to be able to trace the general pattern of climate in the layers of the atmosphere over Antarctica and relate it to the Southern Hemisphere pattern as a whole. But before we view the Antarctic climate in its general context we should first attempt a simple answer to the first question that people ask about the polar regions: 'Why are they so cold?' The explanation is to be found in the contrast between the relatively small amount of incoming solar radiation (technically termed *insolation*) absorbed by the Earth's surface in high latitudes compared with the amount radiated back into space. The contrast is greater during the winters, which are long and dark, than in the summers when there are long periods of perpetual daylight. On midwinter's day the region north of the Arctic Circle (latitude 66° 33′N.) is in total darkness for 24 hours; in contrast the region south of the Antarctic Circle (66° 33′S.) will then be enjoying 24 hours of continuous daylight. At the poles themselves winter and summer will each be of six months' duration. The reason for this lies in the fact that the axis about which the

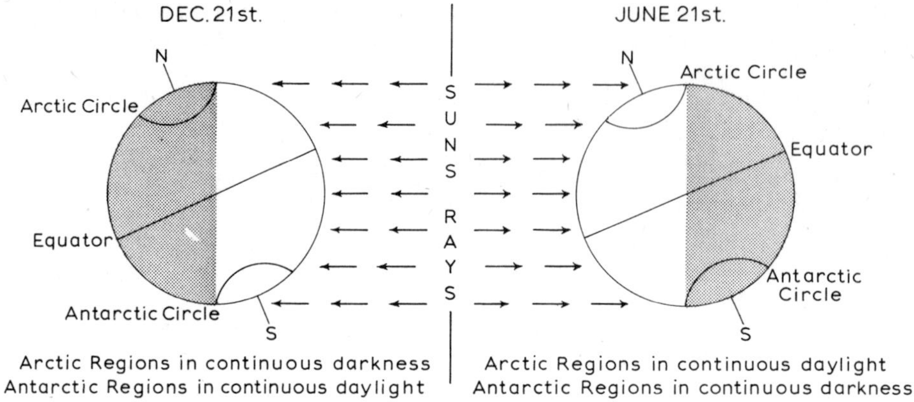

THE POLAR REGIONS AND THEIR RELATION TO DAYLIGHT AND DARKNESS

Earth spins is inclined to the plane of its orbit at an angle of 66° 33′. It is this inclination of its axis which causes the Earth to present to the Sun's rays a larger or smaller area of its Northern and Southern Hemispheres at different times of the year. It is easy to understand why both the Arctic and the Antarctic receive no heat from the Sun during their respective winters. To compensate for this they do receive a large amount of additional insolation in the summer, yet polar summers are on average very cool. One reason is that cloud cover tends to reflect the incoming radiation so that it never reaches the surface. And cloud cover in the Arctic and Antarctic is extensive in summer. A second factor is the reflective characteristic of the Earth's surface which is known as its *albedo*; this varies directly with the colour. The lighter the surface the greater the proportion of radiation reflected. A white sheet of ice, such as that which covers most of Antarctica, will reflect between 50 and 90 per cent of the incoming radiation, and in this way much of the total summer insolation is lost by reflection. Of the very small amount of energy left most is used up in melting the snow surface.

Both polar regions are cold largely for the reasons outlined above. But Antarctica is even colder on average than the Arctic and here the reason must be attributed to its great height. Averaging over 6000 ft. above sea-level it is the world's highest continent and this fact alone, it is estimated, would result in Antarctica being 12 °C. colder than the Arctic. In winter the freezing of the surrounding ocean outward from the coast effectively doubles the area of the continent removing it even further from the warming influence of open water. The Arctic regions, by contrast, consist of a large ocean surrounded by the continental land masses of Eurasia and North America. Though this central Arctic Ocean is covered with a thick lid of floating ice in winter, an underlying current of relatively warm water has a moderating influence on the surrounding coastal zones and prevents temperatures from reaching Antarctic extremes. Indeed, the only comparable low temperatures are recorded on the Greenland ice-sheet and at isolated cold pockets in Siberia and northern Canada.

Antarctica and the Earth's heat budget

Though the Antarctic and the Arctic regions differ greatly both in geography and climate they both serve a vital function in balancing the Earth's heat budget. Very simply this consists of receiving large quantities of surplus heat from the tropics and dispersing it into space. Let us consider in a little more detail how the Antarctic continent functions as a vast radiator of heat. We have already seen that the considerable insolation received by the Antarctic ice-sheet during the summer months is largely reflected back into space and lost. If Antarctica were losing heat in this way year after year one would expect its climate to be getting colder. But this does not appear to be the case. Therefore additional sources of energy must be making up the deficit. What are these sources and from whence do they originate?

One possible source of heat replacement might be an upward flow from the Earth itself. But temperature measurements taken at the South Pole show that while there is a fairly large range of temperatures near the snow surface, at a depth of 30 ft. or so temperatures cease to vary with depth. The loss of heat at the surface cannot therefore be made up by heat from the Earth; it does in fact originate from the Earth's middle latitudes where there is an available heat surplus carried to both the north and south polar regions by *advection*. Advection is the process by which heat is transferred from one point to another by a mass movement of the atmosphere in the form of warm air and water vapour. When this warm air reaches the polar ice-sheets its energy is transferred to the colder surface by a process of direct conduction. If the heat gained by advection exceeds the heat lost by radiation then the surface temperature will rise; if the advective heat is insufficient to compensate for the radiative loss then there will be a fall in surface temperature.

It would seem that this advective inflow of warm air from the lower to the higher latitudes takes place in the lowest layer of the Earth's atmosphere known as the *troposphere*. The troposphere (Greek *tropos*, turning) extends to 6 miles above the Earth's surface above the poles and to 10 miles above the Equator. It is characterized by air temperatures decreasing with increasing height and it is this region which contains most of the water vapour, clouds and storms of the atmosphere. The advected heat in the troposphere is available in two forms; firstly, as sensible heat, that is the heat one can measure with a thermometer, and secondly, latent heat which is the heat released when water vapour is condensed out as a liquid or transformed into a solid such as ice. It has been calculated that 90 per cent of the advective heat transported to the Antarctic is in the form of sensible heat and that the net product of the sensible heat and the latent heat advected balances the heat lost by radiation. We know from the fact that world temperatures vary only very slowly that the radiative heat received by the Earth from the Sun is constant and we can now see how the Arctic and the Antarctic act as *heat sinks* for surplus energy from the lower latitudes maintaining the Earth's heat balance; of these two heat sinks Antarctica is by far the cooler.

Circulation of the atmosphere over Antarctica

The driving force which causes the atmosphere to circulate is, as we have seen, the unequal heating of the Earth's surface. Warm air from low latitudes streams to the Antarctic heat sink but not in a straightforward flow; the motion of the air is controlled by a combination of numerous and complex forces such as differences in pressure and the rotation of the Earth itself. The general circulation of the troposphere in the Antarctic is what is termed 'zonal', that is a regular and constant flow in a clockwise direction round the Antarctic continent parallel to the lines of latitude.

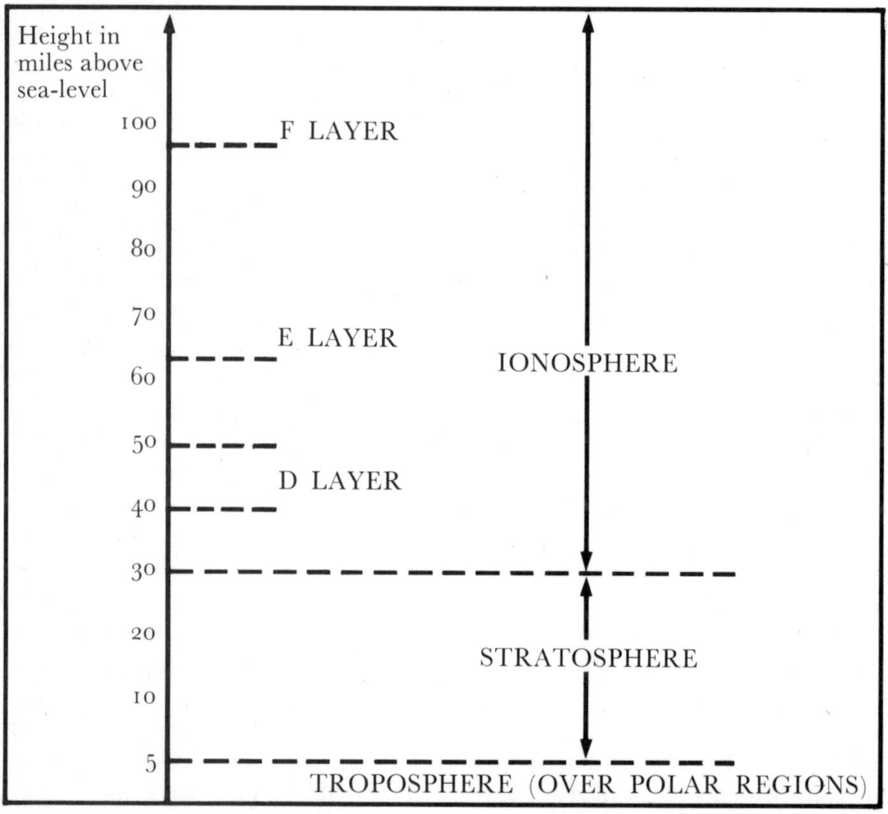

STRUCTURE OF THE ATMOSPHERE

In the Southern Ocean westerly winds predominate from about latitude 40°S. to the Antarctic Circle with easterly or south-easterly winds covering the coastal area of Antarctica and extending inland in some places. The effect of this ring of strong westerly winds is to prevent any sudden exchanges of air with the rest of the Southern Hemisphere; occasionally Antarctic air does break out and may reach as far north as southern Brazil or southern Australia. Here lies an important contrast to the atmospheric circulation in the Northern Hemisphere where in winter cold Arctic air frequently flows out over wide areas of Europe, Asia and America.

Important agents in the transport of air from the ocean to Antarctica, and vice versa, are the storm systems, or cyclones, which form over the Southern Ocean and move eastward towards the continent. Steered by winds in the upper atmosphere these cyclones may bring widespread cloud and snow accompanied by violent winds to some coastal areas. Few of the cyclone (or low-pressure) systems penetrate Antarctica's high plateau though a few will take the short cut across the continent by way of the low-lying ice-sheet which divides the Ross and Weddell Seas.

Antarctic stratosphere

Since the International Geophysical Year of 1957–8 meteorologists, with the aid of balloon- and rocket-borne instruments and earth satellites, have been exploring the stratosphere, that region of the atmosphere above 7 miles up. The stratosphere attains a height of 31 miles above the Earth's surface and is separated from the underlying troposphere by a boundary called the 'tropopause'. Unlike the troposphere the stratosphere is characterized by a slow increase in temperature with increasing height and is a dry and cloudless region. Over the Antarctic continent the temperature pattern of the stratosphere differs in other ways from that on the surface. For example, measurements at the South Pole show that the coldest day of the year on the surface tends to fall a day or two after the reappearance of the Sun. In the stratosphere temperatures appear to rise a few weeks before the Sun rises, reaching a rapid maximum before the Sun has reached its greatest height and then returning to normal summer levels. But this pattern is not invariable – there may be simply a gradual rise to summer levels. Meteorologists attribute this 'explosive stratospheric warming' not merely to the return of the Sun but to variations in the circulation of the stratosphere itself. The cooling of the stratosphere in autumn and winter over Antarctica causes the development of a vast cyclonic system over the Antarctic plateau with strong west winds revolving round a central cold core which increase in intensity with height and spread out as far as latitudes 45° or 50°S. As we have already seen these westerly winds are strong enough to prevent warmer air from the Southern Hemisphere from moving in towards Antarctica. But with the return of spring the westerly winds weaken allowing warm air to flow in. It is this influx

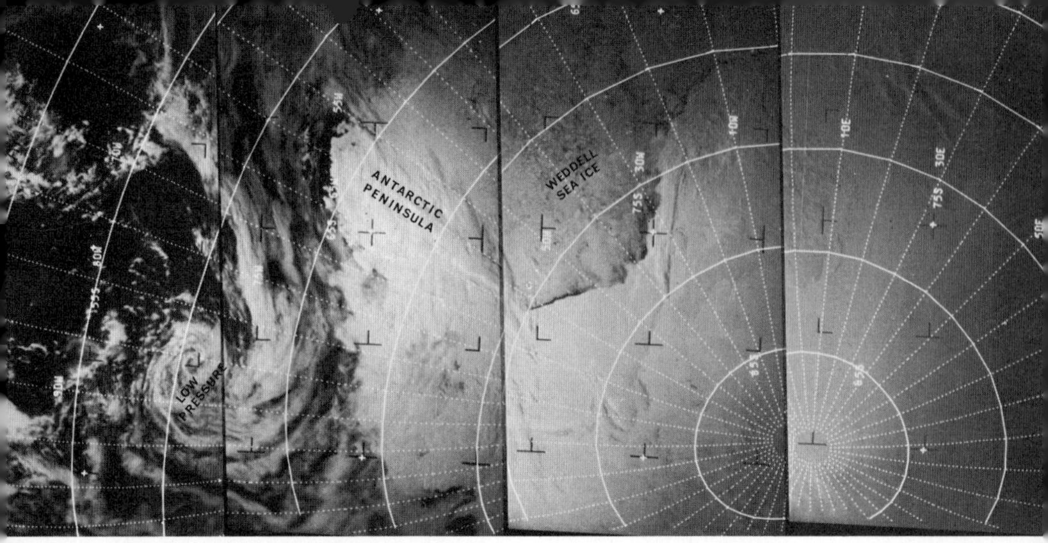

Essa 3 satellite photograph of Antarctica taken from a height of several hundred miles. The whole of the continent is photographed by this means every day

of warm air which contributes greatly to the sudden rise in the spring and summer temperatures over Antarctica.

Ozone

The stratosphere is a region which is characterized by relatively large amounts of the gas ozone, a form of oxygen. Its significance to life on Earth is that it protects us from much of the Sun's harmful ultra-violet radiation by converting it into heat. The process depends on a reaction between the Sun's radiation and the oxygen in the atmosphere; over Antarctica this reaction produces a constant layer of ozone in the lowest layer of the stratosphere; smaller concentrations of ozone are also found at surface level. During the Antarctic winter when the Sun is below the horizon no ozone can be formed, yet curiously enough it is found that ozone concentrations both in the stratosphere and at ground-level reach a maximum at this time. A possible explanation of this seeming paradox is that additional ozone may be generated by static electricity caused by blizzards. Ozone can be a valuable tracer in studying the pattern of the atmospheric circulation of Antarctica.

Carbon dioxide and radioactivity

Two other tracers of value to the climatologists are carbon dioxide and the radioactivity of the air. Antarctica is especially suitable for measuring changes in the amounts of atmospheric carbon dioxide owing to the absence of seasonal variations caused by plant life or man's industrial activities. The natural radioactivity over Antarctica is also low and injections into the upper atmosphere from atomic-bomb tests and other sources can be measured seasonally in the Antarctic stratosphere and are valuable guides to the movement of the stratosphere.

Local weather

Having said something about Antarctica's function as part of the world's weather machine, we turn now to matters of more immediate concern to the explorer and scientist. What sort of weather conditions can one expect to encounter in the field whether on the high inland ice, in low-lying coastal areas or on some off-shore island? The weather of Antarctica can be unpleasant; low temperatures, constant high winds and blinding snow-storms combine to make life at the bottom of the world at best uncomfortable and at worst lethally dangerous to those who fail to take the elementary precautions. We learn from experience, and technology has now provided the means to protect us from the worst the weather can do. But the early explorers, especially those who wintered, suffered greatly and it is to the pages of their diaries that we must turn for accounts of local weather conditions written from the heart. To share, if only vicariously, these discomforts turn to the diary of Captain Scott for those final terrible weeks on the return journey from the South Pole in January–March 1912; or read in *The Worst Journey in the World* what it is like to go hunting for Emperor Penguin eggs in the depth of the Antarctic winter. Here we quote a passage by M. C. Lester from Tom Bagshawe's *Two Men in the Antarctic*; with Lester the author experienced the complete gamut of the Antarctic climate on the coast of Danco Land which is on the west coast of the Antarctic Peninsula (latitude 64° 48′S., longitude 62° 43′W.), the only two-man expedition ever to winter in Antarctica. Though lacking the extremes of the high plateau the climate seems to have been no less uncomfortable.

'Taking a retrospect view of the weather we experienced, my impression, generally speaking, is very bad. At no time during the year could the conditions be called very severe, but they were trying in the extreme. The very great and rapid fluctuations in temperature, with the consequent freezing up and thawing out of everything, the incessant winds and overcast sky accompanied by heavy precipitation and thick weather, were detrimental to our work and would prove trying to the most philosophical of minds. The continual changes in the temperatures gave us little opportunity of becoming acclimatized. For instance, in July and during an interval of only twenty-four hours, a change of wind from the north-east to the south-east quarter brought a temperature of +29 °F. [−1.7 °C.] down to −11.5 °F. [−25.2 °C.] in the course of a single day. Degrees, however, represented little in the matter of personal sensations, and a high temperature of from 25 to 35 °F. [−3.9 to +1.7 °C.] accompanied by a wind and a saturated atmosphere was infinitely more trying than a calm or even a moderately windy day, but with a perfectly dry atmosphere and with a temperature ranging between say +10 °F. [−12.2 °C.] to several degrees below zero; in fact, under the latter conditions existence became cheerful.'

Top left: meteorologist, Deception Island. *Top centre:* releasing a meteorological balloon, Deception Island. *Top right:* radar wind-finder tracks path of balloon-borne instruments. *Right:* anemometer tower, Halley Bay. *Bottom left:* a remote weather station in Mac. Robertson Land. *Bottom right:* meteorologist with radiosonde, a balloon-borne instrument which simultaneously measures pressure, temperature and humidity and transmits the information to a ground based receiver. The balloon can achieve heights of between 80,000 and 100,000 ft. above sea-level

This picture is fairly typical of the 'oceanic' climate of the north-lying Antarctic Peninsula. All parts of the Antarctic south of the Antarctic Convergence are likely to be cold and windy, certainly colder and windier than the corresponding latitudes in the Northern Hemisphere. From day to day, as Lester observed, weather conditions can vary considerably and the weather from year to year may vary greatly also. December and January tend to bring fair weather though even these months are not reliable, as Scott found to his cost.

Temperatures

Indisputably the world's record low temperatures all originate in Antarctica. At the Soviet station of Vostok, over 10,000 ft. above sea-level and near the 'pole of cold', −88 °C. has been measured. Here on the high plateau of Greater Antarctica where the air is especially transparent, due to its low density and freedom from moisture, radiation from the snow surface will largely be reflected back into space. Virtually the whole of this highland region will experience mean annual temperatures of around −55 °C., cold enough to freeze fuel oil into a jelly. The map shows the mean annual isotherms (or lines of equal temperature) for Antarctica over the year as a whole including the mean temperature at some selected stations. From this it is clear that the pattern of temperature over the continent is proportional to latitude and elevation. Round the coast temperatures vary from 0 °C. in midsummer to −20 °C. in midwinter. Only in sheltered places along the west coast of the Antarctic Peninsula will the temperature rise a degree or two above freezing-point in summer. Regular daily variations in temperature tend to be small because of the Sun's low angle of elevation, but surprising variations can be caused by the appearance of low cloud which may raise the surface temperature from 10 to 20 °C. High winds and blizzards usually bring a rise in temperature, though because of the wind-chill effect (see pp. 192–3) an exposed person would actually feel colder. Like other continents, Antarctica experiences a wide annual range of temperature. At the South Pole the temperature varies over 55 °C. during a year. In coastal regions and on the more northerly Antarctic islands the variation is rather less, only 33 °C. in South Georgia, for example.

An unusual feature of the lower atmosphere over Antarctica is the tendency for the air to become warmer at greater heights above ground during the winter months; normally one would expect the temperature to decrease with increasing height. This is because the snow surface is able to lose heat by direct radiation into the clear sky more easily than the air. Air is a poor radiator of heat even when it is quite warm; solid surfaces are usually better radiators. This temperature inversion, as it is called, can result in the air temperature 3000 ft. above the South Pole being 30 °C. higher than at surface level during winter. Even quite close to the snow surface it is possible to detect the increase; temperatures measured at the surface

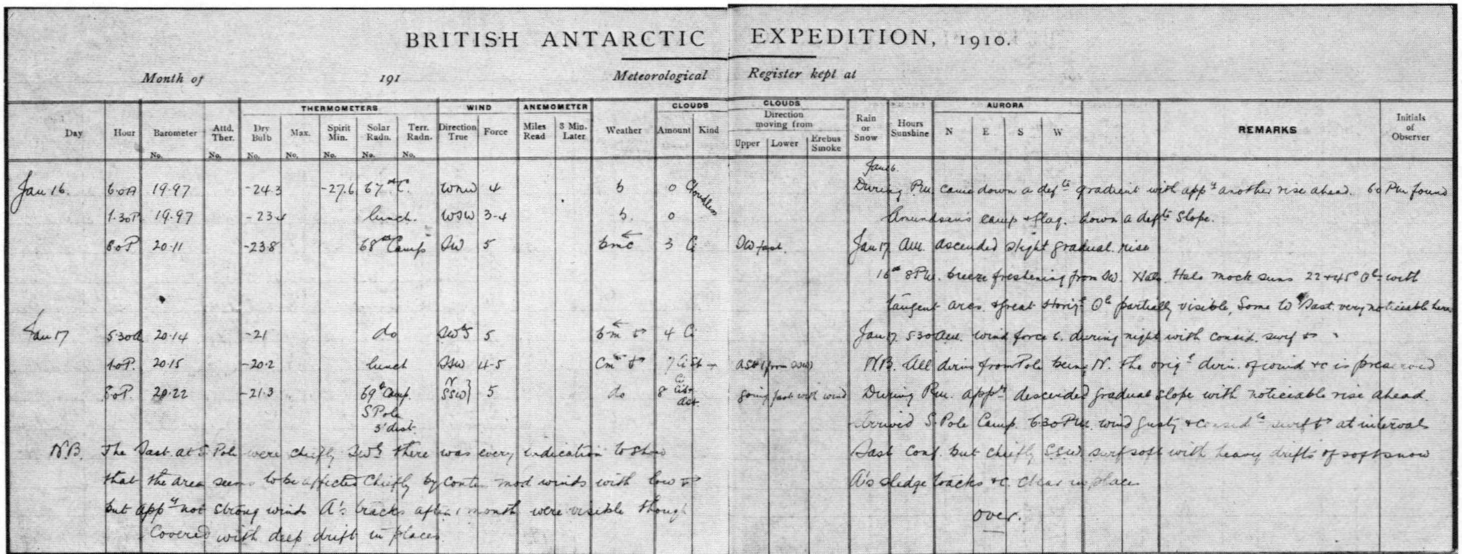

BRITISH ANTARCTIC EXPEDITION, 1910.

Month of 191 Meteorological Register kept at

Day	Hour	Barometer	Attd. Ther.	Dry Bulb	Max.	Spirit Min.	Solar Radn.	Terr. Radn.	Wind Direction True	Wind Force	Miles Read	3 Min. Later	Weather	Clouds Amount	Clouds Kind	Upper	Lower	Erebus Smoke	Rain or Snow	Hours Sunshine	REMARKS
Jan 16.	6.0A	19.97		-24.3		-27.6	67th C.		WNW	4			b	0	Cloudless						Jan 16. During P.M. came down a def[t] gradient with app[y] another rise ahead. 6 P.M. found
	1.30P	19.97		-23.4			lunch.		WSW	3-4			b.	0							Amundsen's camp + flag. down a def[t] slope.
	8.0P	20.11		-23.8			68th Camp		SW	5			bmc	3	Ci	SW fast.					Jan 17 A.M. ascended slight gradual rise
																					16th 8 P.M. breeze freshening from SW. Halo mock suns 22+45° O[c] with
																					tangent arcs + great Horiz[l] O[c] partially visible. Some SW Sast. very noticeable here
Jan 17	5.30A	20.14		-21			do		SWS	5			bm	4	Ci						Jan 17 5.30 A.M. wind force 6 during night with consid. surf dr
	1.0P	20.15		-20.2			lunch		SSW	4-5			Cm dr	7	Ci St	ASt (from SSW)					N.B. All dirn from Pole being N. the orig[l] dirn. of wind &c is preserved
	8.0P	20.22		-21.3			69th Camp. S Pole 3' dist.		N SSW	5			do	8	Ci Ast	going fast with wind					During P.M. app[y] descended gradual slope with noticeable rise ahead. Arrived S Pole Camp 6.30 P.M. wind gusty + consid[l] surf dr at intervals

N.B. The Sast. at S Pole were chiefly SWS there was every indication to show that the area seems to be affected chiefly by Contl. mod winds with low dr but app[y] not strong winds A's tracks after 1 month were visible though covered with deep drift in places

Sast Conf. but chiefly SSW surf soft with heavy drifts of soft snow A's sledge tracks &c. clear in places

over.

Facsimile of the meteorological log of Henry Robertson Bowers, a member of Captain Scott's South Pole party. The log is open at 17 January 1912, the day before the pole was achieved. 'A' in the notes is Amundsen.

are on average 1 °C. below the air temperature 6 ft. above the same surface.

Another curious aspect of Antarctica's climate pattern is that temperatures do not drop steadily with the onset of autumn to rise only with the coming of spring. At any period during the depth of winter, that is between April and August, there may well be one or more temporary rises in the temperature. Like that of the Arctic, the Antarctic winter appears to have no 'bottom' and for this reason is called by climatologists a *kernlose* (German for 'core-less') winter. The simple explanation is that in early winter the heat lost by radiation from the snow surface is quickly replaced by advected warm air from northern latitudes. By late winter the increased effective continental area of ice caused by the freezing of the surrounding ocean helps the radiational cooling to control the situation. In springtime the process begins to reverse with the return of the Sun.

Winds

Along with the record for low temperatures Antarctica also holds that for high winds. Earlier in this chapter we said that the circulation pattern in the Southern Ocean is dominated by westerly winds as far south as the Antarctic Circle. These are the strongest and most constant westerlies found anywhere on the Earth. Round the coast of Antarctica easterly winds predominate and these can be very much influenced by local winds. Many of these local winds are not caused by the general circulation but result from cold dense air rolling down the continental slope from the high Antarctic plateau. Called 'katabatic' (Greek *katabasis*, going down) or gravity winds they are characterized by their strong gusts. Much the windiest part of Antarctica is Terre Adélie. Here in 1912 Sir Douglas Mawson's Australian Antarctic Expedition recorded an average wind speed of 50 m.p.h. (gale force), with gusts exceeding 90 m.p.h. at Cape Denison. Mawson's book *The Home of the Blizzard* contains numerous vivid and entertaining accounts of the hazards of everyday life in the teeth of these persistent winds. To make any sort of progress the art of 'hurricane walking' had to be mastered; this involved lashing crampons to one's boots. Mawson wrote: 'Shod with good spikes, in a steady wind, one had only to push hard to keep a sure footing. It would not be true to say "to keep erect", for equilibrium was maintained by leaning against the wind. In course of time, those whose duties habitually took them out of doors became thorough masters of the art of walking in hurricanes – an accomplishment comparable to skating or ski-ing. Ensconced in the lee of a substantial break-wind, one could leisurely observe the unnatural appearance of others walking about, apparently in imminent peril of falling on their faces.'

Blizzards

It is these strong winds that are responsible for the Antarctic blizzards

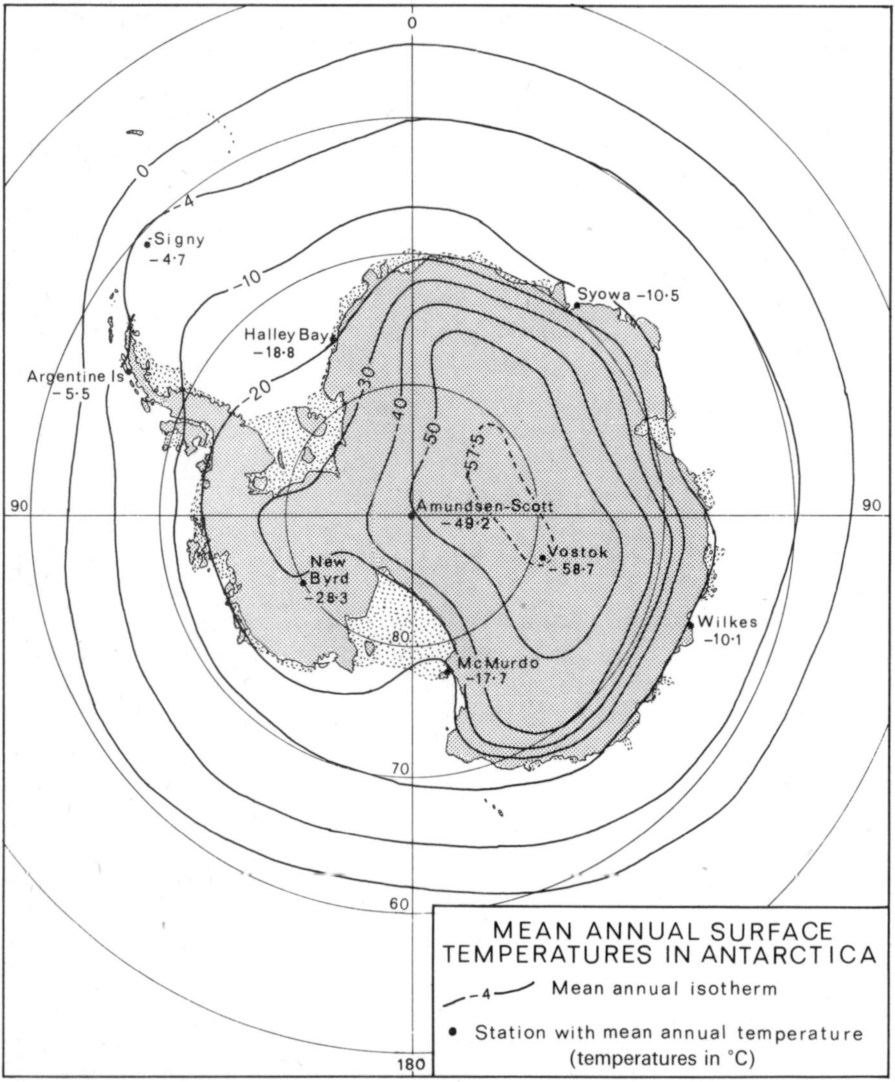
0
90
90
180
60
70
80
0
-4
-10
-20
30
40
50
57·5
Signy
−4·7
Halley Bay
−18·8
Argentine Is
−5·5
Syowa −10·5
Amundsen-Scott
−49·2
Vostok
−58·7
New
Byrd
−28·3
Wilkes
−10·1
McMurdo
−17·7
MEAN ANNUAL SURFACE
TEMPERATURES IN ANTARCTICA
4 Mean annual isotherm
Station with mean annual temperature
(temperatures in °C)

which are in fact not falling snow but surface snow driven along at great force. Blizzards tend to be more common in winter and less so in the summer months. The characteristic sculptured snow-drifts known as *sastrugi* are also caused by the wind's action. Drifting snow is, of course, a major problem to expeditions. This, too, is colourfully described by Mawson: 'Picture drift so dense that daylight comes through dully, though, maybe, the sun shines in a cloudless sky; the drift is hurled, screaming through space, at a hundred miles an hour, and the temperature is below zero Fahrenheit. You have then the bare, rough facts concerning the worst blizzards of Adelie Land.' The actual experience of them is another thing. Shroud the infuriated elements in the darkness of a polar night, and the blizzard is presented in a severer aspect. A plunge into the writhing storm-whirl stamps upon the senses an indelible and awful impression seldom equalled in the whole gamut of natural experience. The world a void, grisly, fierce and appalling. We stumble and struggle through the Stygian gloom; the merciless blast – an incubus of vengeance – stabs, buffets and freezes; the stinging drift blinds and chokes. In its ruthless grip we realized that we are, as Henley says, but

> '*poor windlestraws*
> *On the great, sullen, roaring pool of Time.*'

Along with the Gobi and the Sahara, Antarctica ranks as one of the world's great deserts. The air above the ice-sheet is extremely dry and precipitation consists entirely of fine granular snow whose annual average water equivalent has been estimated at only 2 ins. We have already seen in an earlier chapter how difficult it is to distinguish between falling snow and snow blown by wind so that figures for snowfall can only be calculated indirectly. Small amounts of rain fall in certain coastal areas and along the west coast of the Antarctic Peninsula.

Optical phenomena

No account of the Antarctic environment would be complete without mention of the striking optical phenomena to be observed there. Because of the very dry and dust-free air there is a complete lack of the haze which usually shrouds the view in our own latitudes. Mountains have been sighted at 300 miles distance. This clearness, together with the absence of familiar objects such as trees and houses makes distance judging difficult for the uninitiated. It is disconcerting to discover that an object seemingly only 4 or 5 miles distant is in fact nearly 30. The mirage, familiar to travellers in hot deserts, is also experienced in Antarctica. A whole mountain range may be lifted into the air and objects well out of sight over the horizon loom suddenly into view. The cause is refraction of the light due to the inversion of temperature which we mentioned earlier.

"Solar phenomena on the barrier" by E. A. Wilson

Beautiful and unusual sky effects are created by the myriads of tiny many-faceted ice crystals which float suspended in the dry atmosphere. Light reflected and refracted from these crystals causes mock suns, or parhelia, and mock moons, or parselenae, as well as arcs and haloes round the Sun and Moon themselves.

Visibility

Except near the coast there is very little fog reported from Antarctica; bad visibility is more likely to be caused by blowing snow. There is another hazard which goes under the name of 'whiteout' and is especially dangerous to the unwary explorer; while visibility, in the sense of clarity of the air, can be good, all the contrast, due to the light and shade produced by normal directional illumination is lost so that there is no longer any awareness of rise and fall in the landscape. Under whiteout conditions it is all too easy to stumble into a crevasse or fall over an ice cliff. Whiteout is equally a menace to low-flying aircraft since pilots can be misled into underestimating their altitude above ground. There are a number of meteorological conditions causing whiteout including continuous low cloud, blowing snow and fog.

Our knowledge of the Antarctic climate is far from complete and many of the general statements we have made about it will no doubt need to be modified as fresh information becomes available. We have said enough to indicate that Antarctica with its low average temperatures and great aridity is the Earth's greatest cold desert. Yet Antarctica is an essential component in the world's climatic mechanism, a heat sink for surplus warm air from the tropics and, because of its great height, a vast radiator of cold air influencing much of the Southern Hemisphere climate.

Micrometeorology

There are two associated fields of climatology which we have not so far mentioned, both germane to Antarctic research. These are micrometeorology and palaeoclimatology. Micrometeorology is concerned with the study of the fine structure of the weather at ground-level and immediately above and is especially relevant to the field of plant and animal ecology. It has been observed, for example, that the surface temperature of some isolated, snow-free rock surfaces high up on the Antarctic plateau may rise to as high a level as 28 °C. in the Sun, though the air temperature of the surrounding area may be sub-zero. Such rock surfaces can support lichen life within a few hundred miles of the South Pole and it is clear that while the sun is shining these lichen communities are enjoying a microclimate akin to a warm summer's day under temperate conditions.

Palaeoclimatology

Palaeoclimatology, by contrast, is on the grand scale, the study of climates past wherein we may hope to find evidence as to the nature and sequence of climates yet to come. Here three disciplines – climatology, geology and glaciology – meet together in a common cause. As we saw in a previous chapter, we look to the geological evidence for the first few hundred million years of Antarctica's history when the continent underwent numerous cycles of climate, including an early ice age followed by a period of temperate conditions, as fossil trees bear witness. The reasons for these very long-term fluctuations are still unknown; possibly the continent drifted, possibly the position of the South Pole shifted, possibly it was a combination of both. For the more recent climatic history of Antarctica we have to turn to the evidence provided by glaciology. The deep cores taken from the inland ice-sheet at Byrd Station form a frozen archive which the expert eye can interpret in many ways to provide information on temperature, wind, and the composition of the Earth's atmosphere thousands of years ago. Taking all the available evidence into account it seems that Antarctica has been ice-covered for at least the last 150,000 years. In general its climate seems to have followed the cycle of world climate. The warming of climates 15,000 to 5000 years ago, known as the Post-glacial Climatic Optimum, was shared by Antarctica as was the colder epoch which followed and which culminated between 1000 and 300 B.C. Since then there has been a number of minor cycles of warming and cooling. We do not yet know for certain whether the Earth is heading for a further deterioration in world climate leading to a catastrophic advance of glaciers. With 8 per cent of the total land area of the Southern Hemisphere covered with Antarctic ice we are certainly very close to the last ice age. The final answer to this problem can only be resolved by prolonged studies of Antarctic ice and Antarctic climate.

5

THE UPPER ATMOSPHERE

The Antarctic, as we can now appreciate, forms an ideal vantage-point from which to view certain events taking place in the lower layers of the atmosphere where the world's weather is generated – the troposphere and the stratosphere. We turn now to the upper regions of the atmosphere, 30 to 200 miles above the surface of the Earth. This is the region of the ionosphere, an electrically conducting zone whose behaviour is under the control of the Sun's radiation and where there is a close interaction with the Earth's magnetic field. This interaction of solar and terrestrial events in the ionosphere constitutes the discipline called 'upper atmosphere physics' and is the subject of this chapter.

The upper atmosphere has been intensively investigated in the Antarctic during the past decade, for its characteristics here – and this is true also of the Arctic – are essentially different from those observed in middle latitudes. The geographic poles are close neighbours of the magnetic poles, and it is in these regions, where the Earth's magnetic lines of force converge, that we are best able to observe magnetic storms, auroral displays and certain changes in the composition of the ionosphere responsible for the periodic disruption of radio communications. Only in the polar regions can we observe the effect of prolonged sunshine and darkness on the ionosphere. Antarctica itself is in many ways a most convenient site for upper atmosphere studies. Both the South Magnetic Pole and a zone of maximum auroral activity are located on the continent. Antarctica provides a stable platform for observatories, free from man-made interference and nowadays almost as accessible as many places in the Arctic. In the context of upper atmosphere physics both the Arctic and Antarctic are complementary regions; magnetic disturbances, certain audio-frequency radio signals and auroral displays can occur simultaneously at magnetically conjugate points in both hemispheres.

The inter-relationship between the Sun, the Earth and the upper atmosphere has been compared to the working of a complicated electrical machine. Let us now look more closely at the component parts; firstly at the Sun as a radiator of energy; then at the Earth as a vast magnet; lastly at the ionosphere as an electrical conductor and reflector. Unlike the weather, these remote events on the frontier between the atmosphere and outer space do not affect our senses directly; only the mysterious and transcendentally beautiful aurora are a visible sign of magnetic or ionospheric disturbances. It is with a description of these polar lights that we conclude this chapter.

The Sun

The Sun lies 93 million miles distant from the Earth which travels round it in an elliptical orbit at over 18 miles per second. In essence the Sun is a sphere of gas, heated by thermonuclear reaction at its centre, in which hydrogen is changed to helium. As an end-product of this process of nuclear fusion the Sun emits two kinds of radiation; firstly wave radiation, consisting principally of heat and visible light waves, but also ultra-violet and X-rays, and radio waves. These waves are of differing frequencies, just as musical notes differ in frequency according to pitch, and have varying wavelengths ranging from very short X-rays to very long radio waves. In addition to wave radiation the Sun also emits energy in the form of particles of matter which are constantly ejected from its outer atmosphere or corona. This steady emission of waves and particles, known as 'solar wind', corresponds with the Sun in its quiet state. But at regularly recurring intervals these quiescent periods are interrupted by outbursts of energy which, though they in no way affect the heating properties of the Sun, add to the background radiation and produce violent repercussions in the Earth's atmosphere. These bursts of activity are due to various types of eruption from the Sun's surface, the most powerful being the solar flares, jets of hydrogen gas rising thousands of miles into the solar atmosphere. There are a number of indications which give astronomers and physicists advance warning of these solar disturbances. The most familiar is the appearance on the Sun's face of the dark patches known as 'sunspots', first recorded over 2000 years ago by the Chinese. Sunspots appear to vary in intensity in a cycle of about 11½ years. There have recently been two major programmes of research involving world-wide observations of solar activity and quiescence, both with particular emphasis on investigations in the Antarctic. One, the International Geophysical Year (IGY) of 1957–8, was planned to coincide with a period of maximum sunspot activity and was in the event characterized by an exceptional number of solar flares. The second, the International Years of the Quiet Sun (IQSY) of 1964–5, corresponded, as its name suggests, with a cycle of minimum solar activity. As a result of rocket, satellite and ground-based information from these 'years' we now have a much fuller understanding of the events following a solar flare.

Among the observable effects of solar flares are bursts of ultra-violet radiation and radio 'noise' travelling at about 186,000 miles per second; these are the first emissions to reach the Earth's atmosphere. Taking a more leisurely course are the streams of particles called 'protons', the positively charged nuclei of hydrogen atoms, travelling at between 600 and 1200 miles per second and taking between 20 to 40 hours to make the journey to the Earth's outer atmosphere. Because the protons are electrically charged they will tend to travel down the lines of force of the Earth's magnetic field, entering the atmosphere near the magnetic poles and causing auroral and magnetic disturbances. The approach of these protons may be heralded many hours earlier by the advent of protons accelerated to much higher speeds ranging from 6000 to 60,000 miles per second. Particles in this category are called 'cosmic rays'. Only a small proportion of these are produced by the Sun (solar cosmic rays), the majority originating from outside the solar system (galactic cosmic rays). Very little cosmic radiation is detected at the Earth's surface and that, for the most part, is the debris resulting from collisions with atoms in the upper atmosphere. A few high-speed cosmic rays do succeed in breaking through the Earth's magnetic umbrella and reach the ground in middle latitudes. But it is the low-speed cosmic rays, spiralling down the lines of force that are, for the most part, recorded only in the Antarctic and the Arctic.

The magnetic poles

At this point we need to understand something of the magnetic properties of the Earth. That the iron ore known as 'magnetite' could attract certain metals was known to the Greeks 600 years before the birth of Christ; but the theory that the Earth itself could act as a giant magnet was first mooted in 1600 by William Gilbert of Colchester, physician to Queen Elizabeth I. From Gilbert's treatise, *De magnete*, arose the concept of a magnetic field covering the Earth's surface. The pattern of the lines of force within the field of a simple bar magnet is familiar to most of us from elementary experiments with iron filings. These show the lines apparently emanating from the north pole of the magnet and terminating at the south pole, though in fact they continue within the magnet forming closed loops. In theory the Earth can be considered as having just such a powerful bar magnet buried in its centre with the axis tilted at approximately 11.5° to the axis of rotation. This dipole, if extended, intersects the Earth's surface at latitude 78° 30′N., longitude 69°W. in north-west Greenland, and latitude 78° 30′S., longitude 111°E. in Antarctica, about 600 miles from the corresponding geographical poles. Because these geomagnetic poles are purely hypothetical they do not move and are therefore used as the basis of a system of geomagnetic latitude and longitude. The geomagnetic equator of this system is inclined at 11.5° to the geographic Equator and the zero meridian.

The Earth's magnetic field, like that of the dipole magnet, can be conceived of as an infinite number of lines of force looping thousands of miles out into the upper regions of the atmosphere and converging in the polar regions where they enter the atmosphere's lower layers. The major source of this field stems from an electric current system within the core of the Earth itself. Because of certain irregularities in this little-understood internal current system there are numerous regional anomalies or deviations from the theoretical dipole field. For this reason the actual magnetic poles do not correspond with the theoretical geomagnetic poles. For clarity the magnetic poles are usually referred to as the *dip poles* since at these points the end of a freely suspended compass needle would dip straight down. Unlike their geomagnetic counterparts the magnetic poles tend to drift a few miles each year. The South Magnetic Pole is currently in the coastal regions of George V Land, Antarctica; the North Magnetic Pole in the vicinity of Bathurst Island in the Canadian Arctic. Note that the magnetic poles are not diametrically opposite each other.

It was the British astronomer Edmund Halley who, in 1700, voyaged over the North and South Atlantic Oceans collecting measurements for the first magnetic chart showing curves connecting all places where the magnetic declination had equal values. In the mid nineteenth century the search for the South Magnetic Pole led to important geographical discoveries in Antarctica as we shall see in a later chapter.

So far we have considered the Earth as a semi-permanent magnet whose field is maintained by motions within the Earth's own core and which, though by no means constant, changes only slowly from year to year. In addition to this steady field are smaller contributions from outside the Earth which are characterized by their constant variations from minute to minute and from hour to hour. These changes can be recorded on an instrument called a *magnetograph* which traces magnetic fluctuations in the form of a line on a sheet of photographic paper. This line appears either as a smooth curve or as a series of very irregular fluctuations. Days represented by the smooth curve are called 'magnetically quiet days'; fluctuating lines indicate 'disturbed days', or, if the disturbance is especially violent, 'days of storm'. The amount of variation depends on latitude and on the time of year. Variation is greater in years of sunspot activity, less when the Sun is quiet. The Sun is indeed the indirect cause of these magnetic variations but the mechanism is highly complex. Tides in the upper atmosphere, caused by the Sun's gravitational pull, have the effect of inducing vast electric currents in the ionosphere and it is the magnetic fields of these currents as they move slowly round the Earth which produce the quiet day variations in the Earth's field. Sudden variations and storms, following solar flares, can be attributed to an increase in the electric currents flowing in the ionosphere with a consequent increase of the magnetic field at the Earth's surface. Many magnetic events are believed to be due to charged particles from the Sun being guided down the lines of force and affecting the ionosphere in both the Antarctic and Arctic regions simultaneously. To establish the truth of this hypothesis numerous observations have been carried out in recent years at stations lying at or near magnetically conjugate points, that is at either end of an individual line of force but, because of local disturbances, these points are usually only approximately known. Examples of such stations are sub-Antarctic Macquarie Island and Kotzebue, in Alaska, Little America, on the Ross Ice-Shelf and Baker Lake, Northwest Territories, Canada. There appears to be a good correlation between simultaneous magnetic activity at these stations in the Antarctic and Arctic.

The ionosphere

We have seen that there is a relationship between radiation emitted from the Sun and changes in the Earth's magnetic field, and that the intermediary is an electrically conductive layer of the upper atmosphere called the 'ionosphere'. This region has been the subject of study in recent years, though its history in the Antarctic dates back to Sir Douglas Mawson's wireless experiments in 1911–12. Mawson succeeded in establishing radio communication between his base in Terre Adélie and Tasmania via a relay station on Macquarie Island. Records of the fluctuating strength of signal reception provided the very first evidence from these high latitudes

Ionospherics hut, Halley Bay

of the reflective and absorptive properties of the upper atmosphere. The first serious experiments recording the existence over Antarctica of the radio reflecting layer, then known as the 'Heaviside Layer', were carried out on Admiral Byrd's First Antarctic Expedition of 1928–30. Today continual observations of the ionosphere are being recorded from Antarctic stations as part of a world network. To appreciate the significance of such observations we need to know a little of the cause and the properties of this region of our atmosphere.

The ionosphere is an electrically conductive layer of the upper atmosphere lying between 30 to 200 miles above the Earth's surface. This layer is caused by two agencies; the first and principal is ultra-violet radiation from the Sun; the second, the penetration of the upper atmosphere by charged particles probably originating from the Sun. The effect of this combined bombardment is to knock electrons off the molecules of air, at these heights extremely rarefied. The process is known as 'ionization' – hence the term 'ionosphere'. It is the presence of these free electrons which account for the ionosphere's electrically conducting properties and which, where the electrons are sufficiently concentrated, cause it to act as a reflector. The number of electrons increases fairly uniformly with increasing height; but there are three regions which have distinct reflecting characteristics, termed the D, E and F regions.

Investigations of the ionosphere are made with an instrument called an

Equipment used to probe the ionosphere

The radio-echo aerial, Halley Bay Station, seen at the end of a blizzard, is one of a number used in the Antarctic to study the aurora and variations in the layers of the ionosphere

ionosonde. This transmits short bursts of radio waves into the ionosphere and enables the operator to measure the time taken for the reflected wave to return to the ground. If the speed of the wave is known then the height of the reflecting layer can be calculated. The higher the frequency the greater the concentration of electrons needed to reflect them, thus, by altering the frequency of the transmitter and noting the change in height of reflection, it is possible to estimate the concentration of electrons at different heights. It is the reflective properties of the ionosphere which cause it to act as a world-wide radio mirror reflecting radio waves back to Earth and making short-wave radio communication possible. Ultra-high-frequency waves – those used for transmitting a television picture for example – will not normally be reflected and will be lost into space.

Routine sounding of the ionosphere is constantly in progress. Electron concentrations at different heights are plotted on graphs known as 'ionograms'. These show that the concentration of electrons varies throughout the day, throughout the year and from latitude to latitude. In middle latitudes the electron density is greatest at midday, when the Sun's radiation is producing more electrons, and predictably at a minimum during hours of darkness. One of the more remarkable features of the ionosphere over Antarctica is the persistence of the high concentration of electrons forming the F region during the long night of winter when there can be no direct ionization of this region by the Sun. The probable cause is not ultra-violet wave radiation but a spilling over of particles into the polar regions from the reservoir of trapped radiation known as the 'Van Allen radiation belts'.

Another method of investigating the ionosphere is to make use of the radio waves transmitted every time there is a lightning flash. The electric sparks constituting the flash radiate waves over a wide range of frequencies some of which are so low as to be in the audio-frequency range. An audio-frequency amplifier and a loudspeaker are of course necessary to hear them. These low-frequency waves are able to travel along the lines of force in the Earth's magnetic field oscillating to and fro until they lose their energy and fade out. The most typical of these oscillating signals heard in the Antarctic, and at conjugate points in the Arctic, are the 'whistlers', so called because of their characteristic 'peee-ooo' sound caused by the fact that the higher notes in the signal reach the listener before the lower ones. Since the whistler signals are slowed down by a concentration of electrons this affects the interval of time between the high and the low notes and gives us an audible measure of electron density. Whistlers are therefore valuable tools for probing the effect of the Sun on the ionosphere and the Earth's magnetic field. The polar regions, where magnetic lines of force reach the Earth's surface from the outermost parts of the atmosphere, are the only direct receiving centres.

Like the Earth's magnetic field the ionosphere is also subject to the Sun's periodic outbursts of radiation. These give rise to ionospheric storms closely related to the magnetic storms we have already discussed. The effects of these sudden ionospheric disturbances are complex. In sunlit regions everywhere there is an increase in the electron density in the D and E regions of the ionosphere. In the Antarctic and Arctic latitudes a phenomenon known as 'polar cap absorption' occurs some hours after the outburst of a large solar flare. In effect the lowest D region of the ionosphere ceases to reflect radio waves, absorbing them instead. The result is an interruption or even a complete 'fade out' of high-frequency radio signals, lasting for periods of between two to three days. This means that all radio communication with the outside world ceases and aircraft must be grounded for reasons of safety. Predicting the occurrence of these ionospheric black-outs over the Antarctic is a vital element in the daily routine monitoring.

The aurora

All these happenings in the upper atmosphere may well seem remote to the majority of us; we cannot feel a solar wind or in any way sense fluctuations in the magnetic field or the ionosphere directly. But there is one visible manifestation of the constant bombardment of our atmosphere by solar radiation – the aurora. Known in the Southern Hemisphere as the 'aurora australis', or 'southern lights' and in the Northern Hemisphere as the 'aurora borealis', or 'northern lights', the aurora has fascinated and puzzled man for centuries. Not many people in middle latitudes will be likely to witness a display, for during the Sun's quiet periods the aurora is normally confined to certain regions near the magnetic poles where it can be observed on almost any night. But at times of sunspot activity and the eruption of solar flares a display may be seen in more populated lower latitudes. A network of stations has been established in the Antarctic and elsewhere in the Southern Hemisphere to study the form, colour and position of aurora; complementing this is an even more comprehensive coverage in the Arctic. Auroral studies are not only of absorbing interest in their own right but are a useful means of exploring the ionosphere and its relationship to the Earth's magnetism as well as providing information on the chemical composition of the upper atmosphere.

The aurora is among the most beautiful and spectacular of all natural phenomena. No artist's brush, no camera can hope to capture the elusive nature of its constantly changing form, the purity of its colours or the complexity of its spatial dimensions. The aurora has to be seen to be believed. There are few verbal descriptions that do it justice. Only in the folklore of primitive peoples and sometimes in literature does one find true poetic insight. The Maoris of New Zealand described the southern lights as 'Tahu-Nui-A-Rangi' – the great burning of the sky. Shakespeare's lines in Julius Caesar:

Fierce fiery warriors fought upon the clouds
In ranks and squadrons and right form of war
That drizzled blood upon the capitol

are a graphic word picture of an auroral display as well as a reminder of the superstitious awe such rare events engendered in the hearts of simple folk. But for a first-hand account of an actual display it is hard to better that of John Biscoe from whose journal of a voyage to Antarctica in 1830–2 we now quote:

'The *Aurora Australis* showed the most brilliant appearance, at times rolling itself over our heads in beautiful columns, and then as suddenly forming itself as the unrolled fringe of a curtain, and again suddenly shooting to the form of a serpent, and at times appearing not many yards above us; it was decidedly transacted in our own atmosphere and was without exception the grandest phenomenon of nature of its kind I ever witnessed. At this time we were completely beset with broken ice and although the vessels were in considerable danger in running through it with a smart breeze, which had now sprung up, I could hardly restrain the people from looking at the aurora australis instead of the vessel's course.'

There is not the least bit of overstatement in John Biscoe's account. The aurora, though usually too faint to be seen in day-time, by night can exceed the full moon in brilliance. The pattern is one of constant mutation both in form, motion and colour. A typical display might begin with the appearance of a luminous band or arc stretching across the horizon, green or white in colour. Perhaps after several hours, the whole arc will begin to move, brightening in intensity as it does so, and breaking up into

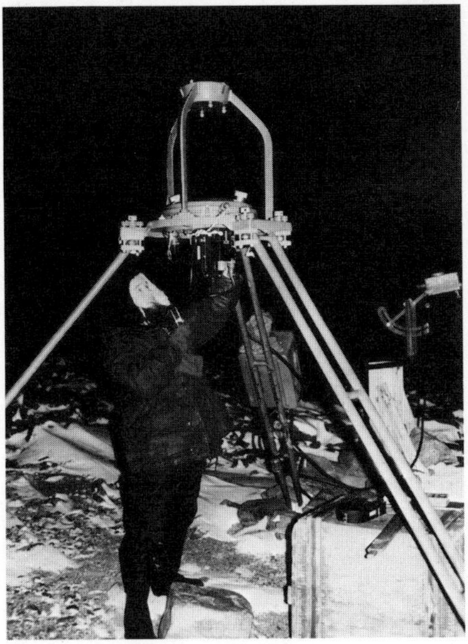

All-sky auroral camera in use during the Antarctic winter at Soviet Novolazarevskaya station

pencils of light of varying length which have the appearance of travelling along the arc with great velocity. This is a *rayed arc*, its colours varying from yellow and red to blue and purple. Finally, long shafts of light focused on one fixed point in the firmament bring the display to a climax with an *auroral corona*. There are in addition numerous variations on the basic arc and rays. Such are the *drapery aurora* in which the whole sky seems to be hung with waving luminescent curtains. Sometimes these are followed by strong waves of red and purplish light moving rapidly upwards, one following the other like a rolling tide of fire; an awe-inspiring sight and known as *flaming aurora*.

Man has observed the aurora from earliest times, but objective study begins with awakening of scientific curiosity following the Renaissance. It was the French philosopher Gassendi in 1621 who, after witnessing displays of these lights as a glow on the northern horizon, named them 'aurora' after the Roman goddess of the dawn. The term 'aurora australis' was coined to describe the southern lights by Johann Reinhold Forster, naturalist with Captain Cook on his circumnavigation of the Antarctic in 1773–5. The first detailed scientific account of an auroral display was given in 1716 by Edmund Halley who noted a connection between the aurora and the Earth's magnetic field. No real progress was made until the First International Polar Year of 1882–3 instituted a co-ordinated programme of synoptic auroral observations from a widely spread number of observatories. A Second International Polar Year in 1932–3 extended the scope of auroral studies and introduced new observational techniques.

Both these events were confined for the most part to the Arctic regions. Not until the International Geophysical Year of 1957–8 did the systematic study of aurora in Antarctica begin in earnest. One of the first auroral stations was that established by the Royal Society of London at Halley Bay.

Until quite recently records of aurora have been based on eye-witness accounts; and there is much to commend this method over all others. But the scale of operations in the Antarctic requiring a continuous record from numerous stations has led to the introduction of more sophisticated techniques, including the camera, television and radar. The all-sky camera can do the work of two men keeping a 24-hour watch during the winter months. The principle is basically simple; an image of the whole sky is reflected in a curved convex mirror; a second mirror reflects this image back to the lens of a ciné-camera through a hole in the centre of the convex mirror. All-sky cameras are usually mounted on a tower to avoid snow drifting, the reflecting mirror being heated to prevent frosting up. A further application of auroral photography enables the height of an aurora to be measured. Here a display is measured simultaneously with two cameras spaced many miles apart. By measuring the apparent shift of the aurora relative to the background of stars the height can be calculated. The lowest part of an aurora may be within 40 miles of the Earth, the upper parts extending skywards for 200 miles or more.

A camera can only operate within the limitations of its lens and the speed of its shutter and film. Many displays are so faint that it is impossible to obtain a useful picture because of the long exposure required and the movement of the subject. A colour film, too, will be slower than a black and white one. One answer to this problem is the television ciné-camera which is so sensitive that it can 'see' and record the faintest and the most rapidly changing of auroral images.

Radar also has its applications in auroral studies. We have already seen that radio pulses can be used to calculate the height of the different layers of the ionosphere. Most auroral forms which occur at night are to be found in the E region of the ionosphere. It is during such displays that ionization of the lower levels of this region occurs, causing radio blackouts over the Antarctic and Arctic. It seems that the auroral forms are associated with local irregularities in the density of the electrons which in turn can be detected by radar echoes. This provides a technique which in certain circumstances enables the aurora to be measured during the day and when obscured by cloud.

It does not follow that you will see a display of aurora wherever you happen to be in the Antarctic. The aurora occurs most frequently within two zones whose boundaries are the lines, where, statistically speaking, one is most likely to witness a display. There is a southern auroral zone, which appears to be approximately circular in form and includes much of Antarctica (though not the peninsula) and part of the surrounding ocean.

And there is a northern auroral zone, elliptical in shape and including the coastal regions of northern Canada, Alaska, Siberia, northern Norway, Iceland and southern Greenland. Auroral displays of a lesser intensity also occur on the poleward side of the auroral zone; called 'polar-cap aurora' these displays tend to take place when auroral zone activity is weak or absent. Following periods of solar disturbance aurora can occur in quite low latitudes. During the International Geophysical Year of 1957–8 brilliant displays were reported in Mexico, Japan and Pakistan as well as in South Africa and Australia.

Earlier in this chapter we noted the relationship between simultaneous magnetic disturbances in both the Northern and Southern Hemispheres due to the 'bouncing' backwards and forwards of solar particles along the Earth's magnetic lines of force. Since the same mechanism can trigger off the aurora it might be reasonable to expect a large-scale display of aurora in the Antarctic to be accompanied by a simultaneous display in the Arctic. Such a relationship between conjugate auroral displays is now definitely established though there are numerous difficulties in the way of demonstrating it. To take two examples; when it is winter and dark in the Antarctic the Arctic is bathed in almost perpetual sunshine; this means that simultaneous observations can only be made in high latitudes during the weeks near the equinoxes. Then there are the difficulties of establishing precise conjugate points and when these are known it can happen that points in the Arctic conjugate to convenient stations in the Antarctic may well be in mid-ocean or in inaccessible or very cloudy regions. The conjugate point to Halley Bay, for instance, is north of Newfoundland, a notoriously foggy region. Improved techniques for observing the aurora in daylight may resolve some of these difficulties.

An answer to the question 'What causes the aurora?' is still puzzling the experts. The Greek philosopher Aristotle attributed them to the Sun's heat causing vapour from the surface of the Earth to rise and collide with the element fire, which then burst into flames. The author of a thirteenth-century Norse manuscript concludes a graphic account of a display of the northern lights as follows: 'Some people maintain that this light is a reflection of the fire which surrounds the seas of the north and south. Others say it is the reflection of the Sun when it is below the horizon. For my part I think it is produced by the ice which radiates at night the light which it has absorbed during the day.' Scientific observation of the aurora began in the seventeenth century alongside studies of the Earth's magnetism. The fact that the direction of a compass needle varied by small amounts during an auroral display was noted in the eighteenth century. During the nineteenth century the quantity of observational data increased greatly as a result of the many expeditions sent to the Arctic and Antarctic. It was discovered that auroral arcs tend to extend along the lines of the Earth's magnetic field. In the opening years of the present century the first

laboratory experiments associating the aurora with electricity and magnetism were carried out by the Norwegian Carl Störmer and an associate Kristian Birkeland. They demonstrated that a high-voltage stream of electrons flowing through a vacuum tube will give rise to a glow similar to that of an aurora. We are all familiar with this effect in neon tubes and in mercury and sodium vapour street lamps with their characteristic blue and orange light. Stormer's experiments reproduced on a small scale the process which takes place in the upper atmosphere when particles collide with molecules or atoms of the air's two chief constituent gases, oxygen and nitrogen. It is the energy dissipated by these collisions, some of which will take the form of light, which is the cause of the aurora's varying colours. Red and yellow are associated with oxygen atoms, red and blue with atoms of nitrogen. A mixture of all colours results in a white auroral light. Information of this kind is obtained by a technique called 'auroral spectroscopy' in which light waves from aurora are broken up into their separate parts and recorded on a photographic plate. The completed record appears as a series of narrow lines (representing radiation from atoms) and bands (representing radiation from molecules). The spectrum can be analysed to give detailed information on a number of properties of the gas in which the aurora is present – temperature and chemical composition for example. It is even possible to detect the fast-moving hydrogen particles entering the atmosphere from outer space and to estimate their speeds – as much as 2000 miles per second.

Many problems will have to be solved before a full explanation of the aurora is available. We cannot, as yet, give adequate reasons for the aurora's constant fluctuations of form, intensity, motion and colour. Most puzzling of all are the complex processes of energy transfer from solar particles to those regions of the Earth's atmosphere where auroral displays are triggered off. We have seen earlier how, during a period of sunspot activity, solar flares increase the normal background radiation. Protons and electrons descend upon the Earth in clouds where they interact with the Earth's magnetic field. Some electrons may be directed down the lines of force to the poles to produce aurora. Other particles may not necessarily be the direct cause of aurora but could have an indirect effect by causing fluctuations in the Earth's magnetic field; and this in turn may shake out electrons stored in those ionospheric reservoirs called the 'Van Allen radiation belts' causing them to enter the upper atmosphere and produce aurora. The theory was put to the test in 1958 when a nuclear device was exploded over the South Atlantic Ocean thereby creating artificial belts of radiation. From this resulted a magnificent aurora which was also observed at a conjugate point near the Azores in the Northern Hemisphere.

\urora': auroral rays, with *Discovery* in winter quarters by E. A. Wilson

6
THE LIFE OF THE ANTARCTIC

'White desert' is an epithet not inappropriate to much of the continent of Antarctica, a place long thought of as a sterile region devoid of all life. By contrast with the Arctic regions, where there is a relative abundance of land animals and over 500 species of flowering plants, there is some truth in this generalization. Antarctica has no large land animals – only a few tiny insects and microscopic animals – and its plants, if we exclude 2 scarce flowering species, are mainly mosses and lichens with a few fungi and algae. These diminutive forms of life are largely restricted to the few regions round the coast which are free of ice and snow in summer, as are the other forms of Antarctic wildlife – the seals, penguins and other marine birds that come ashore to breed. All these animals must rely entirely on the ocean for their food. And here we meet with one of nature's great paradoxes; for Antarctic waters are as rich in life as Antarctic land is poor, supporting a vast population of plants and small animals which in their turn sustain vast numbers of penguins and sea birds as well as marine mammals such as seals and whales.

Scientists have been recording the nature and distribution of Antarctic plants and animals since Captain Cook took the German naturalist Johann Reinhold Forster on his voyage of southern exploration in 1773–5, and Joseph Dalton Hooker, eminent naturalist and botanist, accompanied Sir James Clark Ross to the Antarctic in 1839–43. Since then there have been many studies of the larger animals – the birds, seals and whales – mostly in the Antarctic Peninsula and the Ross Dependency. Rather less is known about the plants and insects. In recent years there has been a tremendous growth of interest in Antarctic biology. Collections of specimens have been made in newly explored regions of the continent as well as on the islands. The range of biological studies has been greatly extended to include many new disciplines. Research today is concentrating on such problems as distribution, that is what species of life exist, where they are to be found and in what numbers, and how they have migrated; on ecology, that is

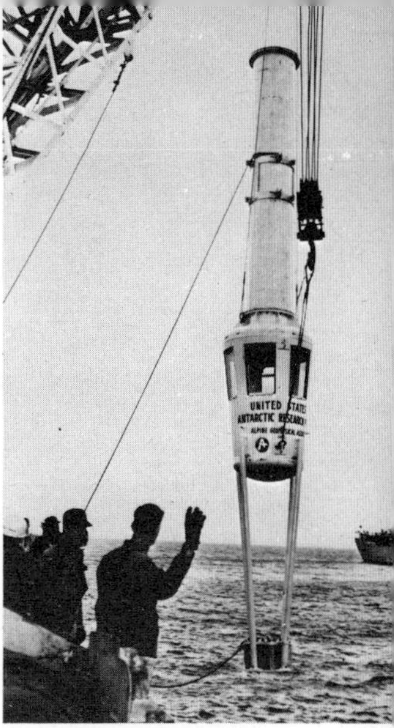

Frogman collecting zoological specimens from a hole in the sea-ice, Signy Island

Observation chamber for use under sea-ice

Marine life trawled from the bottom of the Weddell Sea. It includes brittle stars (*left and right*), starfish (*centre*) with a sea urchin to its left. The many-legged isopods (*2 left, 1 right*) are related to the woodlice

the factors which govern distribution, such as behaviour, food, natural enemies, climate and conditions within the soil or water in which plants and animals exist; and on physiology and anatomy, more especially the way in which plants and animals live and physically adapt to a singularly hard environment.

In this chapter we shall be considering some of the plants and animals associated with the land of Antarctica and then the smaller marine life including fish. Subsequent chapters are devoted to an account of the birds, seals and whales. Additional information on Antarctic life will be found in the chapter dealing with the sub-Antarctic islands.

Conservation

But before considering the life of the Antarctic in more detail a few words concerning its future survival would seem in place. The methods by which Antarctic plants and animals have evolved to survive in an environment to which man himself is not yet fully adapted is a topic of immediate interest. The natural ecological balance of the region is an exceedingly fine one; man-made interference, however unintentional, could easily destroy whole species of animals and some rare plant communities. The presence of semi-permanent scientific stations round the coast of the Antarctic continent and on the Antarctic islands, where vast numbers of seals and birds breed in relatively few isolated places, is a potentially great danger. Fortunately, the possibility of the penguin following the dodo into extinction is now remote. A number of measures for conserving the Antarctic flora and fauna were passed by the Antarctic Treaty Powers in 1964 and these are now being written into the statute books of the various nations concerned. Much further research on the problems of conservation will be needed however before the danger can be said to have passed. A leading expert, Dr. Brian Roberts, put the whole case for conservation most eloquently in a recent paper: 'Let us do our best to preserve for our successors some of the most interesting and exciting biological spectacles to be found anywhere on Earth; the great seal and penguin colonies; the soaring of the Wandering Albatross; the fantastic evening flight of uncountable millions of prions returning to their nesting islands, when the whole surface of the sea appears to be moving against the wind in a solid sheet of gliding petrels as far as the eye can see; a sperm whale leaping clear of the water; a small pink *Colobanthus* flowering in isolated desolation – those who have been fortunate enough to witness these things will never forget the joy they evoked.'

Plants

Many scientific stations in Antarctica maintain an artificial garden of some kind; it may consist of simply mustard and cress on damp flannel or it may be a more complex affair of hardboard, imported soil and fluorescent tubes growing salad vegetables and Dutch bulbs. There is a certain poignancy about this precarious reminder of home and the seeming barrenness of the landscape outside. But the Antarctic desert is a relative one; there is plant life, but one sometimes has to look hard for it. Micro-organisms, such as bacteria, yeasts and the spores of certain moulds are common in some Antarctic soils as well as in the air. Bacteria seem to have remarkable powers of survival; strains have been found existing in sewage dating back to Scott's expedition. The climatic extremes experienced over the continent – prolonged low temperatures, brief and inadequate summer growing seasons, blasting winds and lack of moisture – would be death to the higher forms of plants. Only 2 flowering plants have been recorded on the Antarctic continent; 2 species of the grass *Deschampsia* and a small pearlwort, *Colobanthus quitensis**, all found, though in no great abundance, in the northern half of the Antarctic Peninsula. The sub-Antarctic islands, with their more favourable climatic conditions, can number 30 or so flowering grasses and small flowering plants and ferns. But compare this scarcity of Antarctic plant life with the hundreds of species recorded from the Arctic. The isolation of Antarctica and its surrounding islands from other continents is certainly an important factor.

The most common Antarctic plants are the flowerless cryptogams, a lower order than the phanerogams, or flowering plants. The two main groups are the lichens and mosses and these are widely distributed throughout the ice-free areas of the continent and on the islands of the sub-Antarctic. The lichens form the predominant feature of the vegetation on rocks. Growing on inland peaks within 300 miles of the South Pole, they mark the inner limit of life in Antarctica. Clearly they are the plants best adapted to survive the Antarctic drought. In the region of bird rookeries, where moisture and nourishment derived from guano deposits are relatively plentiful, lichens will be found covering the rock surfaces with patches of yellow or orange. Over 500 species of lichen from the Antarctic are known many of which belong to families distributed throughout the world.

Species of mosses are less abundant than the lichens, because of their greater dependence on water. There are said to be some 70–150 species scattered in coastal areas round Antarctica and a few have been found on nunataks inland. The related liverworts are very scarce; only 6 species have been reported and these mostly from the milder tip of the Antarctic Peninsula.

An unusual sight in the summer months in certain coastal regions of Antarctica are rosy-hued fields of snow. The cause is not a pink precipitation but is due to a species of algae, one of the most abundant plants in the Antarctic. Other species may form blue-green or bright green patches on ice or damp rocks.

* Formerly known as *Colobanthus crassifolius* or Antarctic Pink.

Land animals

Antarctica's dwarfed and infrequent vegetation could hardly be expected to support populations of higher animals and a complete absence of all native land mammals, amphibians, reptiles and freshwater fish is a characteristic of the continent and most of the islands. It is true that there are reindeer breeding on South Georgia and Archipel de Kerguelen, and rats, mice and rabbits on many of the sub-Antarctic islands, but these are creatures which have all been introduced by man – often with disastrous consequences for the native plants and animals. Antarctic land life is the kind that must be studied with the magnifying glass rather than binoculars; it consists entirely of a few small insects and other tiny animals.

Probably the largest group of the insects is the Acarina, more commonly known as 'mites' and 'ticks'. They seem to be particularly resistant to low temperatures and have been found within 250 miles of the South Pole. Some of these insects are able to live independently in mosses in the neighbourhood of penguin rookeries; others find a living as parasites on birds and seals.

One of the most primitive forms of insect life are the springtails or Collembola; these are black, blue or brown in colour, about one-twentieth of an inch in length and always wingless. They occur in the sub-Antarctic and on the Antarctic continent and seem to prefer areas where lichens or mosses are abundant but they also like the underside of rocks where clusters of their eggs are sometimes found.

Not strictly land insects at all are the two orders of lice; they are the Mallophaga, or biting lice, which live among, and chew, the feathers and skin of birds, and the Anoplura, or sucking lice, which live largely on seals.

Antarctica is fortunate in being free of that major bane of Arctic explorers, the biting mosquito, for there are no winged insects on the Antarctic continent; the only native fly *Belgica antarctica* is a small midge found on the west coast of the Antarctic Peninsula. About half an inch long, it is the continent's largest land animal.

Inhabiting the occasional freshwater ponds and patches of damp soil to be found in some coastal regions of Antarctica is a variety of small creatures rich in numbers but few in species. A very primitive species is the one-celled protozoa, the largest of which would just be visible with a hand lens. Next in size are the rotifers, tiny animals which can tinge freshwater ponds pink. The name is derived from the moving cilia round the mouth which give an impression of rotating wheels. Living in close association with the rotifers are the tardigrades or water bears. These are also very small animals with hairy bodies and short spiny legs. Among the dwellers in ponds and damp soils are thread-worms and flat-worms; these are sometimes found in the bodies of seals and birds. Lastly, in a few meltwater pools, may be found Antarctica's only freshwater crustacean, the fairy shrimp.

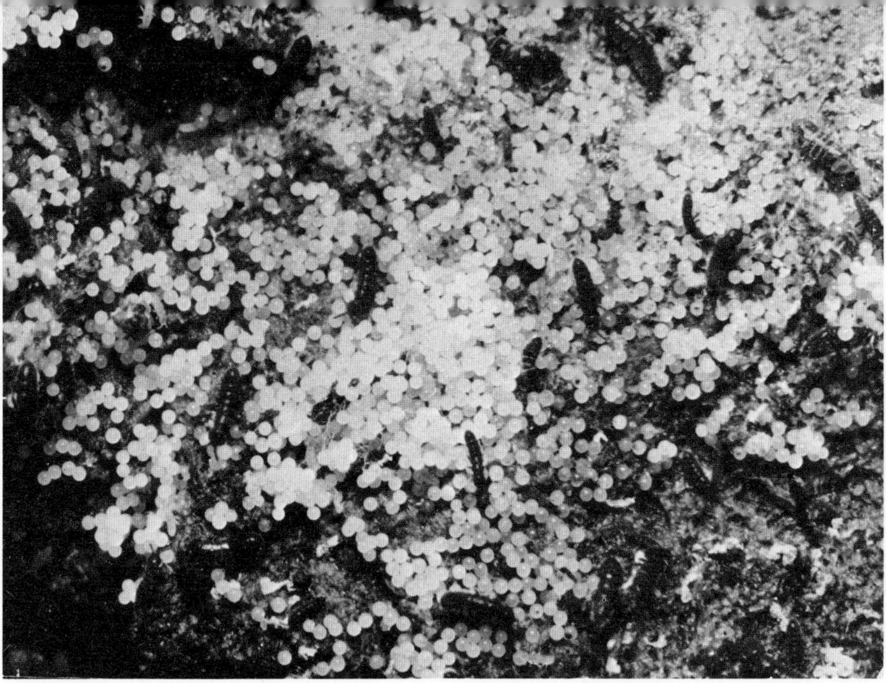

Springtails (*Collembola*) and eggs

A widespread springtail species (*Cryptopygus antarcticus* Willem)

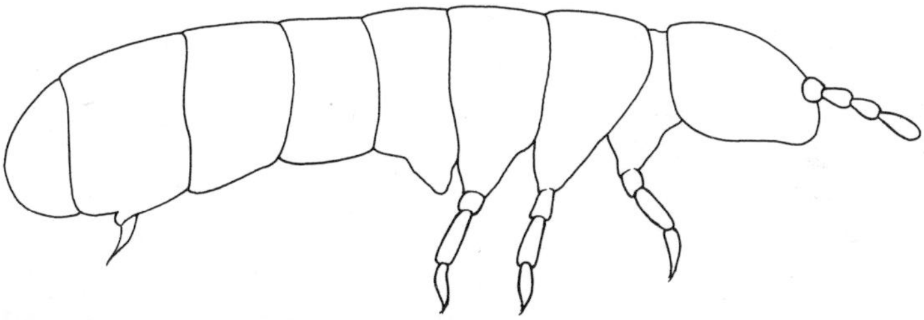

Oribatid mites (*Alaskozetes antarcticus*)

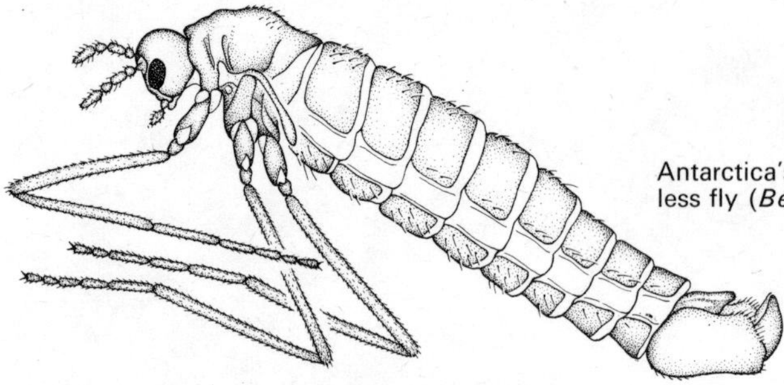

Antarctica's largest land animal, a wingless fly (*Belgica antarctica*)

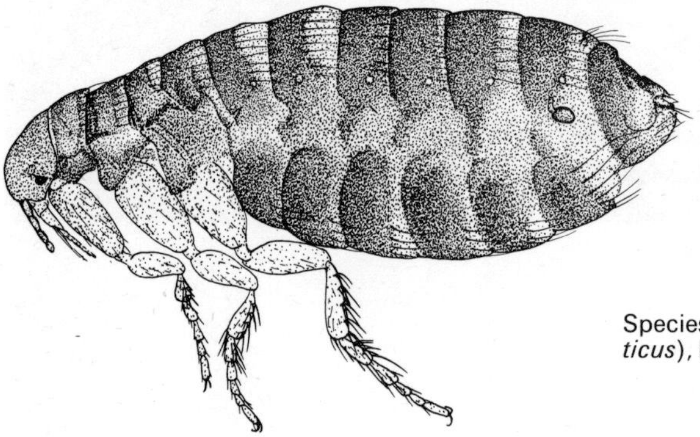

Species of flea (*Glaciopsyllus antarcticus*), believed to be native to Antarctica

These insects and other forms of microscopic animal life provide a remarkable lesson in adaptation. Full use is made of any habitat giving protection from strong winds or providing maximum exposure to the sun. A fairly constant water supply is essential and slopes watered by the melt-streams of snow fields are favoured. Flat stones or crevices in rocks get heated by the sun and often provide the moist warm conditions necessary for life. Mites have been studied on inland nunataks many miles from the coast where temperatures exceed freezing for an hour or two a day during an exceedingly short summer lasting a few weeks only; yet the temperature under the rocks may rise to 20 °C. for short periods. Springtails can withstand temperatures down to about −50 °C. providing the humidity does not become too low. When the temperature drops beyond the point at which metabolism is possible the springtails simply become dormant at whatever stage of development they may happen to have reached, and thus remain until a rise in temperature revives their activity.

One of the most challenging problems yet to be solved in the Antarctic is the origin of these small land animals. The few species represented on the

Entomologists record data on the microclimate of a site inhabited by springtails

continent have relatives on the sub-Antarctic islands and even further afield. For example, the wingless fly, *Belgica antarctica*, is also found in southern South America. Are the present inhabitants of the continent survivors from some ancient temperate Antarctica – a relict fauna? Or are they, perhaps, a more recent migration from surrounding continents and islands, sometime since the last great period of glaciation, the Quaternary? Perhaps both sources are possible, but until knowledge of insect distribution over the Southern Hemisphere as a whole is much more complete the question must go unanswered.

The ocean

In our opening chapter we described a natural ocean boundary dividing the Antarctic from the sub-Antarctic zone, a belt of circumpolar water in which cold northward-flowing Antarctic surface water sinks beneath warmer sub-Antarctic water. This region of mixing oceanic currents is known as the 'Antarctic Convergence'. It roughly follows the 50th parallel of latitude in the Atlantic and Indian ocean sectors but in the Pacific is located between latitude 55° and 62°S. That the Antarctic Convergence

Biologists strain algae from the water of a freshwater lake in Taylor Dry Valley, Victoria Land

is truly an important climatic boundary is the experience of any southward bound traveller on board ship as he crosses the belt. Not only will he detect a marked drop in temperature but he will also observe that the birds of the air and creatures of the sea have changed too; for the convergence is an important boundary for fauna as well as climate.

One inhabitant of these cold Antarctic waters is the red shrimp-like crustacean whose Latin name is *Euphausia superba* but which is perhaps better known as 'krill', the name given it by the Norwegian whalers. Though there are many species of euphausids, krill is found only to the south of the Antarctic Convergence where it forms the basic diet of fish, winged sea birds, penguins, seals and whales. The krill, in turn, subsists on tiny one-celled diatoms, the preponderant plant life of Antarctic waters. This small plant and animal life is known collectively to biologists as plankton, the term 'phytoplankton' being given to the plants and 'zooplankton' to the animals. Its surface distribution is very much at the mercy of the currents so that it tends to drift about in great patches. The zooplankton has the ability to rise to the surface at night and sink during the hours of daylight. Like all forms of Antarctic life plankton is poor in species but rich in individuals, so much so that the Southern Ocean is potentially the richest source of marine protein in the world – a fact that has not escaped the notice of nutritionists anticipating the needs of an expanding world population. The reason for this bounty is the richness of the Southern Ocean in such nutrients as nitrates, carbonates, phosphates and silica; aided by the prolonged sunshine of the Antarctic summer the diatoms transform these elements into living tissue. This basic food chain of diatom-krill-larger marine animals is completed by the return of organic material, in the form of excreta, and dead bodies, to the sea where it is broken down by bacterial action into basic chemicals and thereby made available to recommence the cycle. A jingle by Griffith Taylor in the *South Polar Times* for 1911 sums the whole business up most memorably:

Big floes have little floes all around about 'em,
And all the yellow diatoms couldn't do without 'em.
Forty million shrimplets feed upon the latter,
And they *make the penguins and the seals and whales*
Much fatter.

There are also many other chains in which small fish or squids form the basic food of larger fish, birds, seals and whales. The Southern Ocean is, therefore, far more than mere salt water; it is a rich soup which nourishes a relatively few species of animals but ensures that they are numerically very plentiful.

Benthos

In addition to the wealth of life on the surface there is an abundance of plants and animals on the sea-bottom, collectively referred to as 'benthos'. A large part of this benthos is made up of sponges with needle-sharp spicules. These lowly animals are more abundant in the waters round Antarctica than in tropical oceans. Although sponges are usually considered to have been at their most plentiful in the Cretaceous period, 135 to 70 million years ago, there are some biologists who consider that their true period of ascendancy is in the Antarctic today.

Another common group of sea-bottom creatures is formed by the echinoderms (sea-urchins and starfish). A class of these, the ophiuroids, forms a significant part of Antarctic benthos. These curious star-shaped animals move by pushing and pulling upon surrounding objects with their arms. Other creatures include corals and sea-anemones, jellyfish – one measuring about 1 ft. across – and various marine worms. There are also molluscs, or shellfish, though these are rather restricted in variety. They include mussels, limpets, cuttlefish, snails, octopuses and squids. Marine biologists are paying a great deal of attention to the distribution and

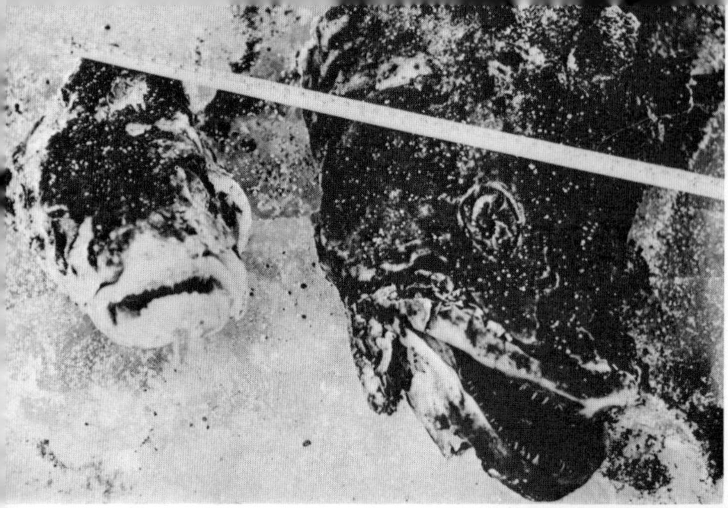

Fish found frozen into the McMurdo Ice-Shelf

Marine biologist with Nototheniiform fish (*Trematomus bernachii*). Studies of fish metabolism and physiology form part of current research on Antarctic biology

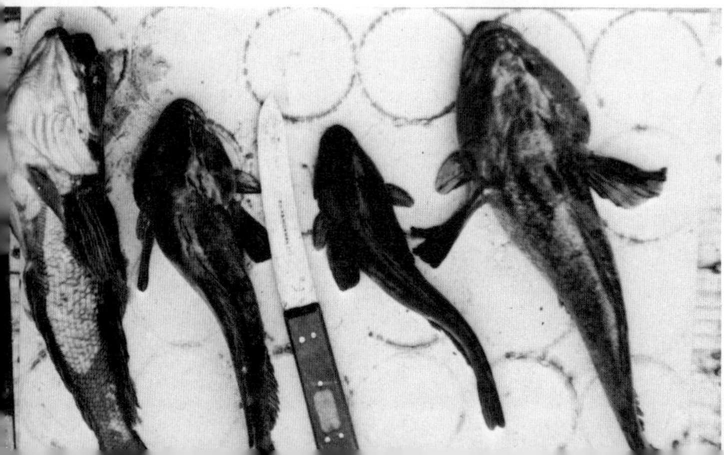

Specimens of Nototheniiform fish

quantity of benthos found in the shallow waters round Antarctica. It is probable that these creatures play an important part in converting waste matter and plankton organisms into food for fish. Underwater photography, submarine observation chambers which can be lowered through holes in the ice, aqualung diving as well as conventional trawling of specimens, are all techniques in current use which enable these and other marine creatures to be studied alive in their natural environment.

Fish

Antarctic fish are typically few in species; of the 20,000 known species of fish in the world only about 100 have been listed for the Southern Ocean south of the Antarctic Convergence. And of these 90 per cent belong to a single group, the 'Nototheniiform fishes', only one being peculiar to these waters. This group consists of four families – the Bathydraconidae (Antarctic dragon-fish), the Harpagiferidae (plunder-fish), the Nototheniidae (Antarctic cod) and the Chaenichthyidae (ice-fish). These fish differ very much in appearance; the first two families, as their names suggest, are somewhat forbidding. The majority live on the bottom or in moderately deep water and few are surface swimmers. Most are edible. One of the largest species, *Notothenia rossii*, common round South Georgia, attains a length of 3 ft. The Chaenichthyidae include members that have no circulating red blood; some of the Nototheniidae are also partially bloodless. Since haemoglobin, the red pigment in blood, is normally needed to carry life-giving oxygen from the gills to the body cells, the metabolism and oxygen transport of these fish is a subject of considerable interest.

A further puzzle is the method by which these fish survive the temperatures of Antarctic waters which surprisingly enough remain extremely stable throughout the year. Though their body fluids contain dissolved salts which lower the freezing-point these fish seem able to survive temperatures that are well below the freezing-point of their blood. Presumably some process of supercooling* must take place. It appears that physical contact with sea ice can cause death by freezing and this may well account for the fact that there are few surface-swimming fish in Antarctic waters.

* Supercooling – the cooling of a liquid below its freezing-point without solidification.

7
BIRDS

Compared with the 120 or so species of birds breeding in the Arctic the number breeding on or south of the Antarctic Convergence and round the continent of Antarctica is very small – less than 50. Only 16 are recorded as breeding in Antarctica itself. But as elsewhere in the region bird life makes up in number what it lacks in variety; Antarctica and its islands are renowned for their enormous rookeries of sea birds. The relatively high survival rate is a measure of their perfect adaptation to what to us appears as a highly uncongenial environment. Natural enemies are also largely absent. Birds will be found only where there is exposed ground where they can nest, and where there is food. Bare surfaces free of ice are, as we can now appreciate, comparatively rare in Antarctica and there are many stretches of coastline where bare rock of any kind is lacking. It is interesting to observe the intensive use made of the little living room that is available. Flat open spaces may be occupied by Giant Petrels with smaller petrels burrowing in the ground beneath them. Other petrels will make use of patches of lichen and moss and every available ledge and niche will be occupied by still other birds.

The chief source of food is naturally the sea which in the summer months supplies Crustacea, squid or fish on or near the surface as well as the usual superabundance of plankton life. In certain regions upwelling of nutrients from the ocean-bottom can greatly increase the supply of food. With the onset of winter the food supply dwindles with the formation of sea ice and the disappearance of the Sun. In consequence most birds must migrate during the winter months, only the Emperor Penguin remaining to breed on the fast ice. Antarctic birds are almost entirely marine. There is no shallow inter-tidal zone which could support waders and inland there is no living thing which can support bird life. During the summer months many gulls and sheathbills scavenge for their living among the refuse of nests of penguins and other birds.

It is hardly surprising that there is an almost complete absence of land birds in the Antarctic. The only true land bird is a pipit living on South Georgia and there are species of duck on South Georgia, Archipel de Kerguelen, Iles Crozet and Macquarie Island.

The marine birds of the Antarctic belong to these five main groups: penguins; petrels; skuas, gulls and terns; shags; and sheathbills. Despite the large numbers of these birds only a few have been studied in any detail.

Penguins

Penguins are the most truly Antarctic of birds; they number 18 species, all breeding in the Southern Hemisphere. Eleven species breed in the Antarctic and sub-Antarctic and of these 4 species breed on the Antarctic continent or on the surrounding pack ice. Penguins have received more attention from visiting biologists than the other birds; they are flightless and therefore easy to capture. One of the best known and most interesting of this group is the Emperor Penguin. It is the largest member of the penguin family, the male bird weighing up to 90 lbs. and measuring up to 3½ ft. in height. When Edward Wilson, Captain Scott's zoologist, visited the famous rookery at Cape Crozier in 1903 it was the only one known. Today many Emperor rookeries have been recorded all round the coast of the continent. Of all Antarctic creatures the Emperor Penguin is the most thoroughly adapted to the severe climate. This adaptation is best seen in its nesting habits. Unlike other birds which nest in the spring in time to launch their young before the onset of winter and the disappearance of food, the Emperor reaches the coast in March just when the ice is beginning to form. In May, which in the Antarctic is the depth of winter with 24 hours of total darkness, the female Emperor lays her only egg on the fast ice which provides a convenient flat surface. Her duties are then taken over by the male partner while she leaves the colony to forage for food at sea after a fast of two months. The male now has an equivalent spell of duty facing him without food; he must keep the egg at a temperature over 30 °C. before it can hatch. This means he must live on his fat which indeed accounts for one-quarter of his body weight at this time. As there are no nesting materials available the father has to carry the egg, weighing about 1 lb., on his feet nesting it in a warm pouch of feathers. In this way he can walk around with the egg and avoid drifting up with snow. By the time the mother returns father will have lost up to 45 per cent of his weight. Emperors have an ingenious way of conserving their fat reserves and minimizing heat loss; they get into a huddle, at first in small groups and then, as conditions grow worse, into one vast group which may attain 6000 strong. New-born Emperor chicks are engaging creatures with their silvery-grey down, which as it lengthens, has the appearance of fine fur. A chick spends the first two months of its life sitting alternatively on the feet of the father or mother who feed it with regurgitated fish, squid or

Common name	*Specific name*	*Breeding areas*
Emperor Penguin	*Aptenodytes forsteri*	Coast of Antarctica and southern half of Antarctic Peninsula
King Penguin	*Aptenodytes patagonica*	Macquarie Island, Prince Edward Island, Heard Island, Iles Crozet, Archipel de Kerguelen, South Georgia, South Sandwich Islands
Gentoo Penguin	*Pygoscelis papua*	Northern half of Antarctic Peninsula, South Shetland Islands, South Orkney Islands, Falkland Islands, Macquarie Island and probably other sub-Antarctic Islands
Adélie Penguin	*Pygoscelis adéliae*	Coast of Antarctica and Antarctic Peninsula, South Orkney Islands, South Shetland Islands, South Sandwich Islands, Bouvetøya, Peter I Øy
Ringed (or Chinstrap) Penguin	*Pygoscelis antarctica*	Northern half of Antarctic Peninsula, South Shetland Islands, South Orkney Islands, South Sandwich Islands, Bouvetøya, Peter I Øy
Rockhopper Penguin	*Eudyptes crestata*	Falkland Islands, Tristan da Cunha group, Ile St. Paul and Ile Amsterdam, Prince Edward Island, Iles Crozet, Archipel de Kerguelen, Auckland Islands, Campbell Island
Macaroni Penguin	*Eudyptes chrysolophus*	South Shetland Islands, South Sandwich Islands, South Georgia, Bouvetøya, Falkland Islands, Iles Crozet, Archipel de Kerguelen, Prince Edward Island, Heard Island
Royal Penguin	*Eudyptes schlegeli*	Macquarie Island
Erect-crested Penguin	*Eudyptes atratus*	Antipodes Island, Campbell Island
Magellan (or Jackass) Penguin	*Spheniscus magellanicus*	Falkland Islands
Yellow-eyed Penguin	*Megadyptes antipodes*	Campbell Island, Auckland Islands

ANTARCTIC AND SUB-ANTARCTIC PENGUINS

krill. It takes six months for an Emperor chick to reach the stage when it can go down to the sea and feed itself. And the chances of survival within this early stage are not great. A chick mortality of 77 per cent has been estimated. Quite apart from the extremes of climate and the fragility of the eggs there are occasions when chickless adults attempt to 'adopt' strays and, in the resulting brawl, the chick itself gets trampled to death. Petrels who kill them for their stomach contents are another hazard.

Most numerous and most popular with visitors to Antarctica (especially photographers) is the little Adélie Penguin. Standing very upright, about 1½ ft. high, he resembles with his shimmering white front and black back and shoulders, a perky little waiter in evening dress. In contrast with the ponderous Emperor, the Adélie averages between 9 and 11 lbs. in weight. But what he lacks in size he more than makes up in speed. Adélies can average up to 3 m.p.h. over the ice either waddling on their legs or tobogganing on their breasts.

The habits of Adélies have attracted the attention of biologists ever since they were first studied by Edward Wilson and G. Murray Levick on Captain Scott's two expeditions. They are keen competitors with scientists for the best camping sites, the rather limited areas of exposed rock round the coast of Antarctica where they live during the summer months in vast rookeries. Here they build their nests of small stones, lay their eggs, raise their families, and depart for the pack ice to the north where they spend the winter months. Their annual cycle is thus the opposite of the Emperor.

The behaviour of these birds in their crowded rookeries – a 55-acre site, it has been calculated, might support a quarter of a million birds – has been much studied in recent years. The first arrivals, usually males, reach the breeding-grounds about the middle of October either singly or in trains, follow-my-leader fashion. Individuals then make directly for last year's nests, which, once found, are protected from intruders with bill and flipper. Each householder then gathers together a pile of stones for a nest – pilfering from his neighbours if need be – and from time to time indulging in what biologists term 'ecstatic displays' – a triumphant stretching upwards of the head accompanied by a slow backwards and forwards waving of the flippers, while at the same time emitting a pulsating note which rises to a climactic and ear-splitting 'aah, aah, aah'. This raucous note attracts the attention of the female of the species who approaches the nest and gives the tenant 'the glad eye' – that is she quite literally stares at him perhaps from one eye only or even squintingly from both. The male replies in turn to the female by staring in a like manner and making a bow over his shoulder. He then lies down and scrapes a hollow in the pile of stones he has gathered. No just cause or impediment now remains to separate the pair from a season of matrimonial bliss – unless of course last year's wife turns up. If she does a pitched battle between the two females inevitably results in the eviction of the husband-stealer and a victory for the principle

Adélie Penguin and chick

Emperors – note the tobogganing position

Emperor Penguins and young

1 Tabular iceberg

2 Royal Research Ship *John Biscoe* in fast ice. Coronation Island in background

3 Measuring ice movement with stakes on the Liv Glacier, Victoria Land

4 The Ferrar Glacier, Victoria Land

5 Snocat of the Trans-Antarctic Expedition 1955–8 in a crevasse

6 Ice 'cave' formed by a snow-bridged melt-water channel. The horizontal lines indicate former water levels

7 Drill in use on the Ross Ice-shelf for extracting ice cores

8 Looking south from Adelaide Island

9 Aerial view of Lemaire Channel, west coast of the Antarctic Peninsula

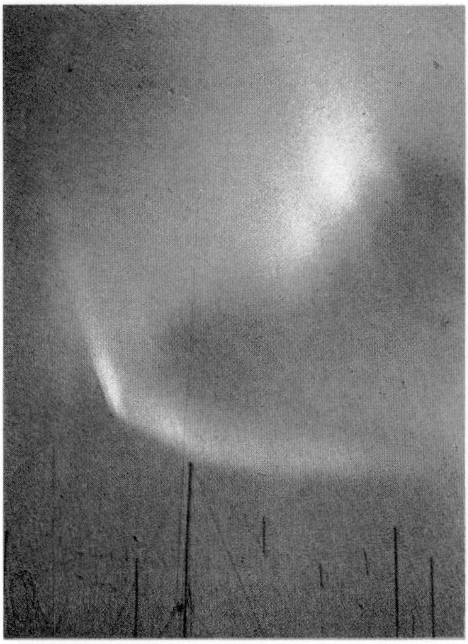

10 Folded auroral band with rays – drapery aurora over Halley Bay against a background of stars

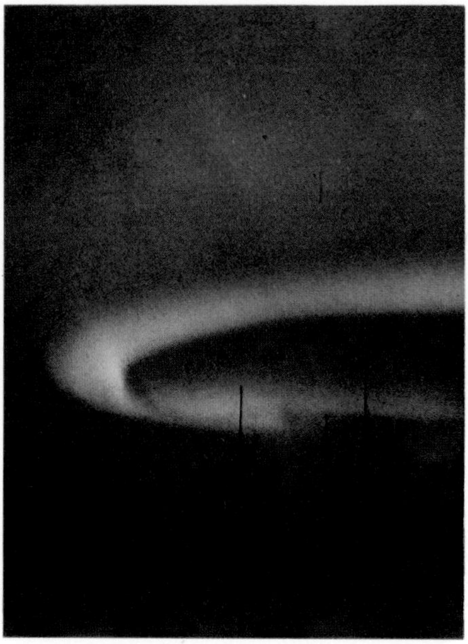

11 Auroral band over Halley Bay

12 Solar halo and parhelia

13 American automatic weather station designed to be dropped by parachute from an aircraft and to transmit weather observations by radio for a period of three months

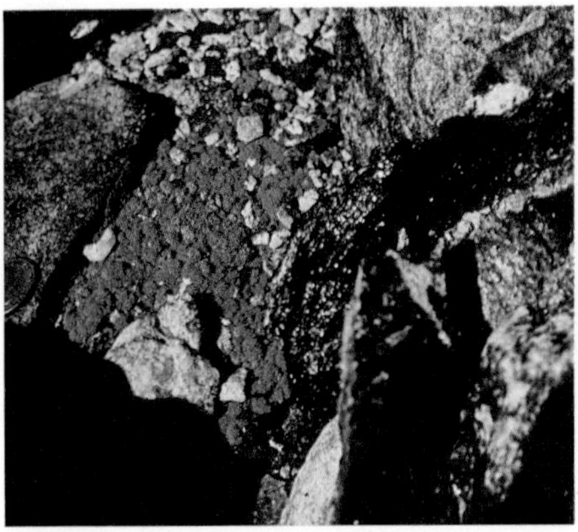

14 Patches of orange-coloured lichen growing near a bird rookery

15 Antarctic grass *Deschampsia antarctica*, South Georgia

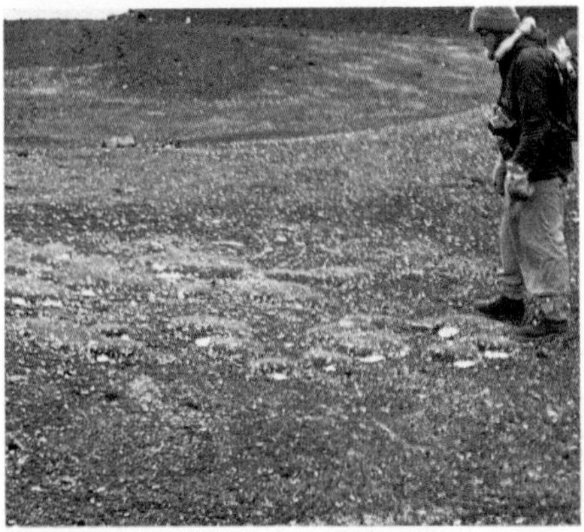

16 Patch of *Deschampsia antarctica* growing in volcanic ash, Candlemas Island, South Sandwich Islands

17 Cushion of Antarctic Pearlwort *Colobanthus quitensis*, South Georgia

18 *Polytrichum juniperinum* and other mosses growing on a moist rocky slope, Signy Island

19 Moss *Polytrichum alpinum*, South Georgia

20, 21 Fruiting heads of Greater Burnet *Acaena adscendens*, a creeping plant which with mosses, such as *Tortula robusta* (*below*) covers acres of scree slopes in South Georgia

22 Antarctic Bedstraw *Galium antarcticum*, found among the tundra grasses of South Georgia and the Falkland Islands. The flowers are only about 2 millimetres in diameter

23 Adélie Penguins at Cape Adare

25 Ringed Penguin

24 Emperor Penguins, Cape Crozier, Ross Island

26 Gentoo Penguin and chicks

27 Sheathbill, Archipel de Kerguelen

28 Black-browed Albatross, Archipel de Kerguelen

29 Courtship of the Wandering Albatross

30 Shags on nest, Avian Island, off Adelaide Island, Antarctic Peninsula

31 Cutting up Baleen Whales

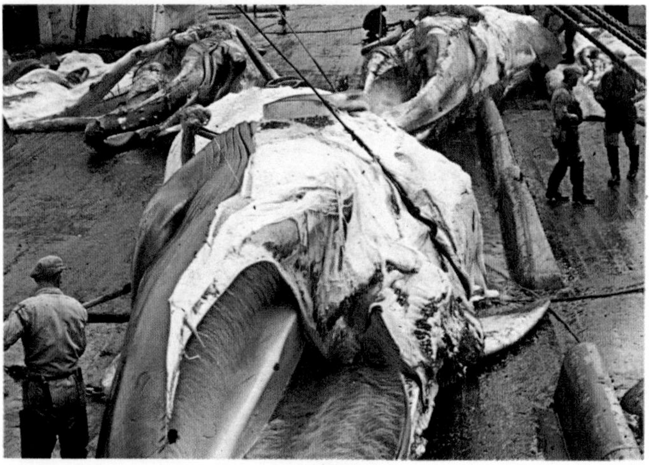

32 Flensing – the process of stripping blubber from the whale

33 Krill

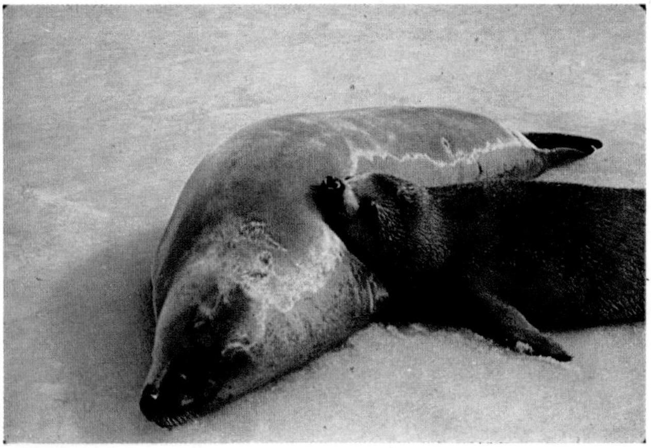

34 Crabeater Seal and pup

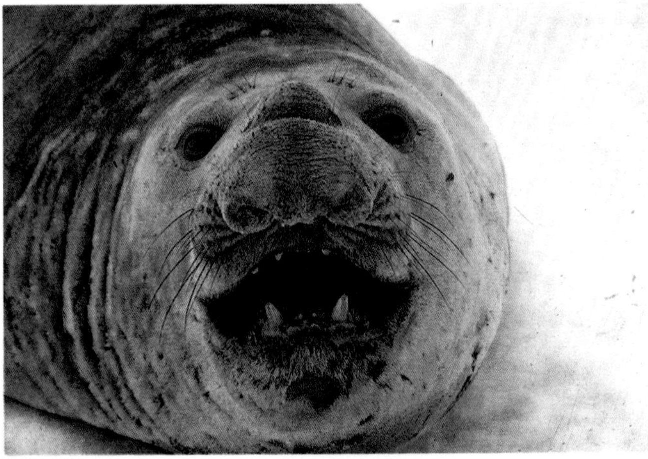

35 Elephant Seal

36 Weddell Seal

37 Stanley, capital of the Falkland Islands, seen from the harbour entrance

38 The former whaling station at Grytviken, South Georgia

39 A summer landscape near Leith, South Georgia. Higher plants, such as *Deschampsia antarctica*, cover the boggy valley floor around the small tarn. Mosses and lichens are abundant on the upper slopes

40 Sunset over the South Orkney Islands. Coronation Island seen from Signy Island

41 A warm oasis in a cold desert. View inside the crater of volcanic Bellingshausen Island, South Sandwich Islands. Note the green of the moss mats on heated ground

42 Tristan da Cunha. A volcanic eruption in October 1961 brought a lava stream to within a few hundred yards of the settlement

43 (*Top*) Penguin rookery, Iles Crozet

44 (*Right*) Baie du Navire, Iles Crozet, with Elephant Seals in foreground

45 (*Below*) King Penguins, Iles Crozet

46 Station Port-aux-Français, Archipel de Kerguelen

47 Kerguelen Cabbage *Pringlea antiscorbutica*. Valued by the sealers for warding off the scurvy, its thick fleshy leaves, when boiled, are not unpleasant to eat

48 U.S. icebreaker *Southwind* in the pack ice off McMurdo Station

49 Royal Research Ship *John Biscoe* at the ice-edge south of the Argentine Islands, Antarctic Peninsula

50 Soviet ice-strengthened freighter *'Ob* with Lisitsin 2 aircraft off Molodezhnaya

51 Danish ice-strengthened cargo vessel *Kista Dan* unloading at Halley Bay

52 U.S. Navy Super-Constellation aircraft at McMurdo Station beneath Mount Erebus

53 Unloading cargo from a Soviet Ilyushin 12 at Vostok

54 Otter aircraft of the British Antarctic Survey at Adelaide Island

55 U.S. Army helicopter at Mount Hope on the Beardmore Glacier

56 Snocat of the U.S. Antarctic Research Program with Weasel tracked carrier alongside

57 Muskeg tractor of the British Antarctic Survey

58 Soviet Chelyabinsk S-100 tractor with sledge

59 Lansing Snowmobile of the British Antarctic Survey at Halley Bay

60 McMurdo Station viewed from Observation Hill

61 Scott Base, McMurdo Sound

62 Fossil Bluff, a British field station on Alexander Island above George VI Sound, Antarctic Peninsula

63 British station at Stonington Island, Marguerite Bay, Antarctic Peninsula

64 Soviet station Novolazarevskaya, Dronning Maud Land

Incidents during the Antarctic voyage of Sir James Clark Ross to the Ross Sea, 1841–2, painted by John Edward Davis, second master of H.M.S. *Terror*

65a Planting the British flag on Possession Island off the coast of Victoria Land, 11 January 1841

65b New Year's Day 1842. A carnival on the ice-floes in latitude 66° 32′S., longitude 156° 28′W.

65c Watering H.M.S. *Terror* in the pack ice, 1842

Adélie Penguins, Graham Land, Antarctic Peninsula

Adélie Penguins on nest of stones

The study of penguin physiology and adaptation is a major aspect of biological research in Antarctica. Here a geologist records data from numbered subjects

Rockhopper Penguins and chick

of monogamy. Just how penguins recognize each other and their nests is a mystery; it seems that they can recognize each other's voices more readily than each other's faces. Adélies taken on air trips 2400 miles from their rookery have been able to find their way back to their own nests.

Egg-laying begins in the middle of November. Two eggs are laid a day or two apart and the male Adélie, like his Emperor counterpart, stays behind to incubate them, while the female takes a couple of weeks off to break her fast. After an incubation period of 34 days the chicks are hatched, guarded by the parents, and fed on a nourishing diet of regurgitated krill. Adélie chicks consume vast quantities of food. At four weeks they weigh over 3 lbs. and can consume this weight in krill daily. It is now the second or third week in January and the chicks are deserted by their parents. They form themselves into protective groups, or crèches, where they huddle until their parents return with food. Even in the apparent anonymity of the crèche, chicks are still able to recognize their own parents' voices. In early February the chicks begin to lose their down and come into their juvenile plumage. They now make for the beach where they remain in groups excitedly watching the older birds swimming. Then they too take the plunge. Once they have acquired the necessary skill, the young Adélies head for the distant pack ice to the north where they remain until they reach breeding age.

The penguins are an ancient race. Fossil penguin species have been found in the Antarctic Peninsula dating from the Lower Eocene period, that is some 60 million years ago. Some of these fossils were $5\frac{1}{2}$ ft. tall, nearly man sized. Penguin embryos display keel bones and wing quills which suggest that they evolved from birds that could fly. An absence of natural predators perhaps reduced the need for flight.

The penguin's peculiar adaptability to water suggests a strong link with the sea birds. Awkward on land, they are superbly efficient under water. Their flippers, vestigial remains of wings, give them a propelling force which enables them to achieve speeds of up to 25 m.p.h. To regain the ice they eject themselves from the water like rockets and soar upwards landing on the ice in an upright position with their flippers at their sides. Speed and agility in the water are essential to survival for the coastal waters round the Antarctic continent are the home of the Killer Whale and Leopard Seal, the penguin's arch-enemies.

Other penguins breeding on the continent of Antarctica are the Gentoo Penguin and the Ringed Penguin. A very numerous species found on some sub-Antarctic islands and as far north as Australia is the Royal Penguin. There was at one time a flourishing penguin oil industry on Macquarie Island where thousands of these birds along with another sub-Antarctic species, the King Penguin, were boiled down for their oil every season. Two other penguin species to be seen in the sub-Antarctic are the Macaroni Penguin and the Rockhopper Penguin.

Gentoo Penguin and chick

Ringed (or Chinstrap) Penguin and chicks

Petrels

There are about 27 different species of petrel breeding within the Antarctic regions including 4 species of albatross. Best known among these is the great Wandering Albatross, with a wing span exceeding 11 ft. and weighing from 17 to 21 lbs. This bird breeds only once every 2 years and the chick takes up to 9 months to fledge. The Wandering Albatross breeds on most of the sub-Antarctic islands but the only species to breed south of latitude 60° is the Sooty Albatross.

Among the petrels nesting principally in this area is the Snow Petrel and Wilson's Petrel. The former is a truly Antarctic bird; an all-white species it is usually associated with the pack ice but in summer returns to the continent to nest in burrows under the rocks. The black and white Wilson's Petrel is the smallest of the Antarctic petrels. It is one of the most abundant birds in the world and migrates annually to the Northern Hemisphere. Well known as a scavenger in Antarctic waters is the Giant Petrel which weighs about 10 lbs. and has a wing span of 7 ft. These birds are great wanderers during their first year. Nestlings banded at Wilkes Station,

Cape Pigeon and egg

Silver-grey Petrels courting

Sooty Albatross on nest

Silver-grey Petrel	*Fulmarus glacialoides*
Cape Pigeon	*Daption capensis*
Antarctic Whalebird or Dove Prion	*Pachyptila desolata*
Snow Petrel	*Pagodroma nivea*
Antarctic Petrel	*Thalassoica antarctica*
Wilson's Petrel	*Oceanites oceanicus*
Blue-eyed Shag	*Phalacrocorax atriceps*
Dominican Gull	*Larus dominicanus*
Antarctic Tern	*Sterna vittala*
Brown Skua	*Catharacta skua* subsp.
Giant Petrel	*Macronectes giganteus*
Sheathbill	*Chionis alba*

BIRDS OTHER THAN PENGUINS RECORDED AS BREEDING IN ANTARCTICA

Antarctica, have been recovered from as far away as Easter Island in the Pacific, South Africa and South America. Despite this wide scattering it seems probable that many of these young find their way home to breed.

Skuas, gulls and terns

The Brown Skua is the most Antarctic of all the southern birds and ranges over the continent to within 120 miles of the geographic South Pole. It is a notorious predator, as we have seen, ready to pounce on a penguin egg or peck out the eyes of a chick; the refuse-bins of scientific stations have provided an alternative source of food in recent years. The Brown Skua is found all round the continent and winters no further north than the pack ice, though stragglers have been located as far north as Australia and India. Skuas mate for life and have a strong homing instinct, regularly returning to the same nesting areas.

Of the 4 terns found in the Antarctic, 2 are of special interest. The Antarctic Tern is a truly Antarctic bird nesting on many of the sub-Antarctic islands as well as on the continent. The Arctic Tern, by contrast, migrates southward from the north polar regions at the end of summer to spend a second summer season in the Antarctic feasting off the krill of the pack ice. Then, at the end of the southern summer, it begins its 11,000 mile journey north again. Thus, with the exception of a few weeks on the wing, this bird spends much of its life in perpetual daylight.

Shags

Three cormorants or shags occur within the region. Only the Blue-eyed Shag breeds on the continent, the remainder being restricted to certain sub-Antarctic islands.

Sheathbills

The Sheathbills, or Kelp Pigeons, complete our list of Antarctic birds. During the summer months Sheathbills are prominent scavengers in the neighbourhood of penguin rookeries where they feed on eggs, chicks and droppings. The Sheathbill is the only bird without webbed toes to breed in the Antarctic.

Female skua sits on an unhatched egg with a newly hatched chick nestling beside her

8

WHALES AND WHALING

Whales are animals with a world-wide distribution, but it was in the krill-rich waters bordering the Antarctic continent that, until recent years at least, they were to be found in their greatest numbers. Whaling has for long been the Antarctic's only commercially profitable industry but over-exploitation has reduced it greatly in importance.

There are many types of whale found in Antarctic waters ranging in size from the commercially valuable Blue Whale, measuring over 90 ft. in length, to insignificant dolphins a mere 10 ft. or so long. Despite their aquatic environment, whales are mammals; they bring forth their young (called 'calves') alive and suckle them with a rich fatty milk. Like other mammals the whale is warm-blooded and has a four-chambered heart. It breathes air through blowholes into its lungs and must hold its breath under water expelling it by the characteristic 'blow'; this is not a spout of water (as so often depicted in old prints) but a visible cloud of oily vapour. There are two main orders of whales (or Cetacea as the zoologists prefer to call them), the Odontoceti, which have teeth, and the Mysticeti – the Baleen or Whalebone Whales – from whose upper jaws is suspended a sieve of horny baleen plates.

Baleen Whales

Most of the largest and commercially important whales belong to this second group. Baleen Whales feed almost exclusively on the shoals of shrimp-like krill which carpet the waters south of the Antarctic Convergence during the months of summer. These they consume by swallowing great mouthfuls of water which they force out of the sides of their mouths with their great muscular tongues past the baleen plates, which taper off into fine hairs. The krill is caught in this hairy mat and it is estimated that up to 2 tons of krill and other small creatures may be consumed during the course of one day by a single Blue Whale.

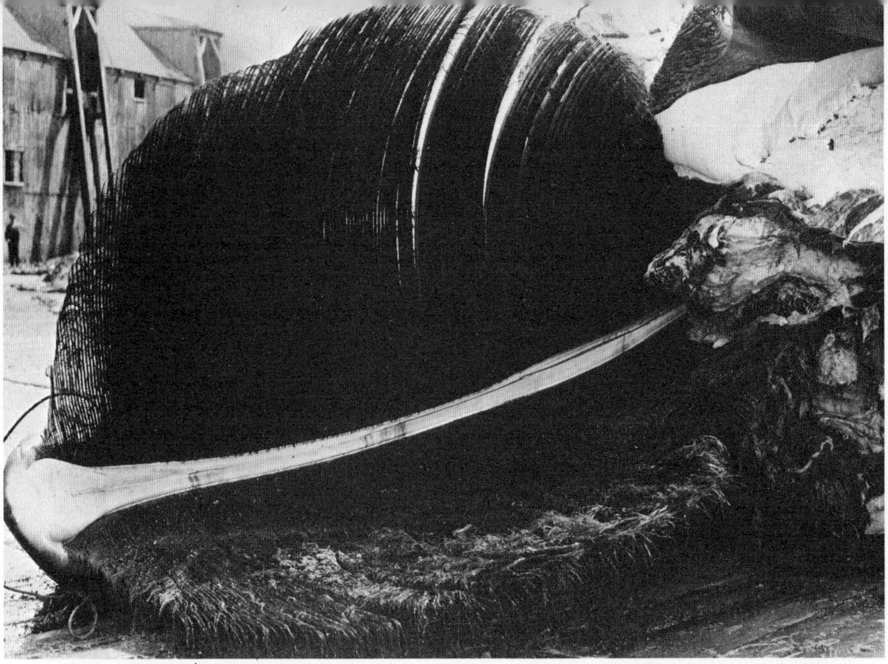

The filter in the mouth of a Baleen Whale

Among the Baleen Whales the largest group is the rorquals. These are characterized by the presence of a series of longitudinal pleats on the chest and throat, by short, broad baleen plates and by a small dorsal fin. First among the 6 species of rorquals found in the Antarctic is the Blue Whale; this is by far the largest of the Cetacea and indeed of all animals alive and extinct. Individuals measuring 96 ft. long and weighing over 140 tons have been recorded. Over a quarter of this weight may be oil and it is therefore hardly surprising that the species has been over-hunted by whalers and is now in consequence very rare. The Blue Whale is slaty blue-grey in colour and tends to frequent the edge of the pack ice sometimes taking refuge in open leads where whale-catchers dare not venture.

Similar in distribution and habits is the Fin Whale. Slimmer in proportions than the Blue Whale it averages 70 ft. in length and 50 tons in

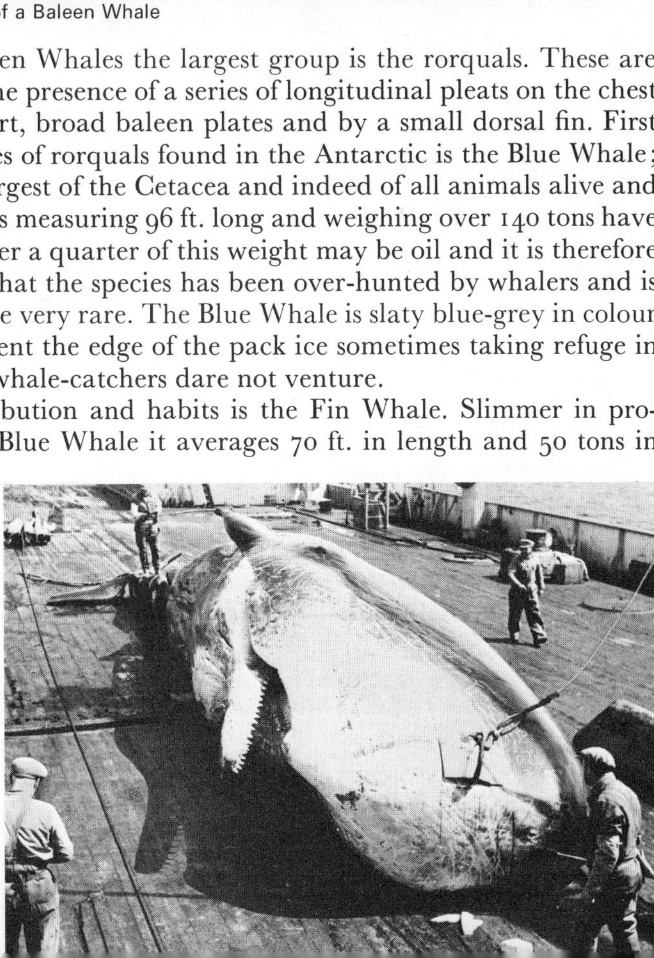

Sperm Whale showing lower jaw armed with formidable teeth

BALEEN (or WHALEBONE) WHALES (Mysticeti)	
Fin Whale or Fin Back or Common Rorqual	*Balaenoptera physalus*
Blue Whale or Sibbald's Rorqual	*Balaenoptera musculus*
Bryde's Whale	*Balaenoptera edeni*
Sei Whale	*Balaenoptera borealis*
Lesser Rorqual or Minke Whale	*Balaenoptera acutorostrata*
Humpback Whale	*Megaptera nodosa*
Southern Right Whale	*Balaena australis*
TOOTHED WHALES (Odontoceti)	
Sperm Whale or Cachalot	*Physeter catodon*
Bottle-nosed or beaked whales (Ziphiidae), e.g. Southern Bottle-nosed Whale	*Hyperoodon planifrons*
Dolphins and Porpoises (Delphinidae)	
Killer Whale	*Orcinus orca*
Southern Dolphins, e.g. Cruciger Dolphin	*Lagenorhynchus cruciger*
Pilot Whale or Blackfish	*Globicephala melaena*

WHALES (CETACEA) FOUND IN ANTARCTIC WATERS

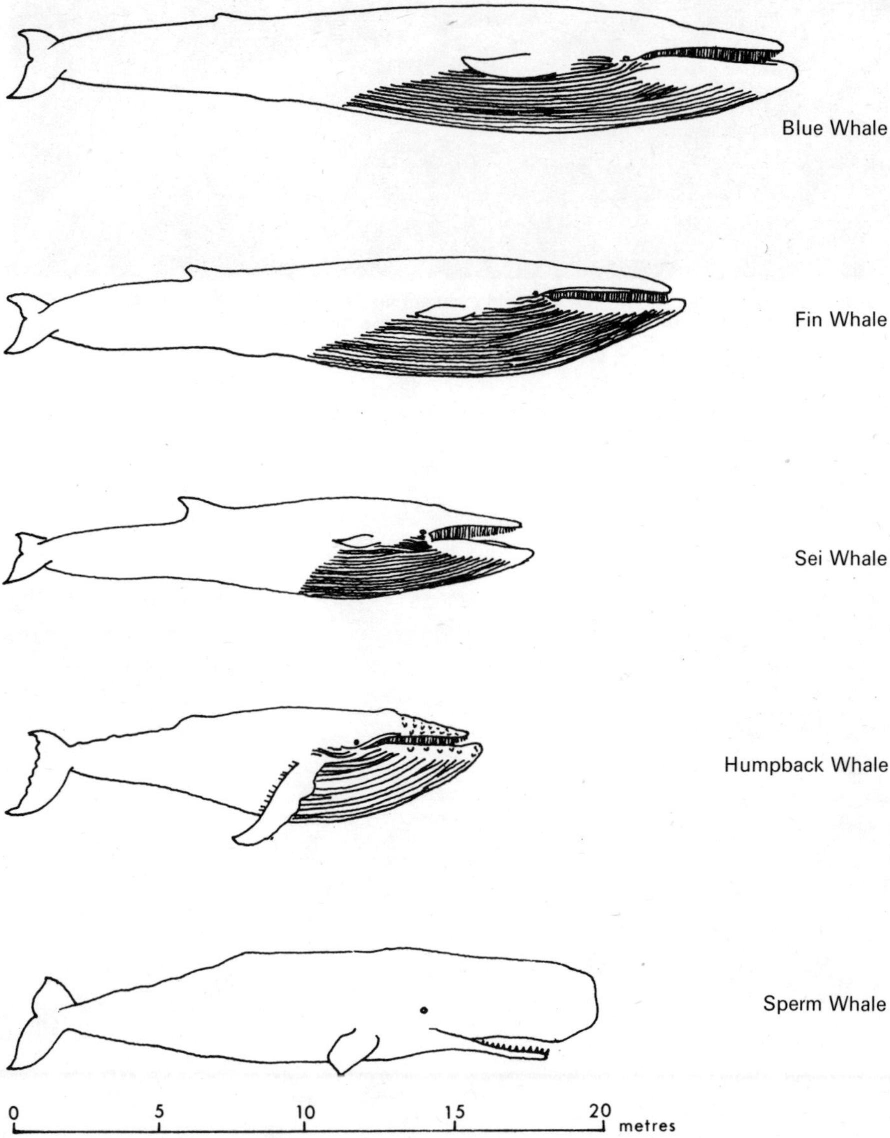

Relative sizes of some Antarctic whales
(From 'The Antarctic Pilot', 3rd Edition 1961)

Euphausia suberba, the shrimp-like krill which swarm in the waters south of the Antarctic Convergence and are the main diet of the Baleen Whale. They range from less than $\frac{1}{2}$ in. to nearly $2\frac{1}{2}$ in. in length

Krill spilling from the open stomach of a 72-ft. Blue Whale

weight. Though the proportion of oil is smaller than that of the Blue Whale, the Fin is now the most plentiful and most sought after of the larger Antarctic whales. It can be distinguished by its white underbelly and by the baleen which is half dark and half cream in colour. Its back fin too is larger than that of the Blue Whale – hence its name.

The Sei Whale and Bryde's Whale are closely related species with only small differences between them; the baleen plates of Bryde's Whale are coarser than those of the Sei Whale probably because the former feeds on fish while the latter mostly takes crustaceans. Both species weigh under 20 tons but are being hunted in increasing numbers.

The smallest of the rorquals, the Minke (pronounced 'minky') Whale can be recognized by the white band across the flipper not present in other whales. In length it measures around 30 ft. and weighs up to 7 tons. The meat, when properly cooked, is said to be indistinguishable from beef, but the Minke is too small in size to be of real commercial importance.

Quite unlike the rorquals is the Humpback Whale. It is relatively short and fat averaging 40 ft. in length and weighing about 30 tons. Characteristic are the exceptionally long flippers. Decorating the head will be found a variety of knobs and lumps themselves often tipped with living barnacles which give the creature a most grotesque appearance. Over one-third of its weight may be oil.

A very rare, and now protected, whale is the Southern Right Whale. Excessive hunting very nearly extinguished this slow-swimming species, extremely rich in oil, but it has been recorded in coastal regions of Cape Province, South Africa, and may be recovering its numbers.

Baleen Whales are to be found during the Antarctic summer in a zone about 200 to 300 miles wide bounded to the southward by the edge of the pack ice and extending round the whole of the Southern Ocean.

Toothed Whales

Less important commercially are the Odontoceti, or Toothed Whales. Instead of baleen plates they have teeth in one or both jaws and their food consists largely of fish or squid. The only Toothed Whale of commercial importance in the Antarctic is the Sperm Whale, or Cachalot, immortalized in fiction as Herman Melville's 'Moby Dick'. Antarctic specimens are invariably males; breeding pods (or 'harems') of bulls and cows are only found in tropical and temperate waters. A male Sperm Whale averages about 50 ft. in length and weighs between 30 and 35 tons. Characteristic of the Sperm Whale are the huge head which makes up about one-third of the total length, and the slender lower jaw, with its two rows of formidable teeth, used to seize and kill squid. Some of these squid themselves attain sizeable proportions; a specimen of the Giant Squid, measuring 16 ft. in length and weighing 405 lbs., was found whole in the stomach of one Sperm Whale. The head and jaws of the Sperm Whale are usually scarred by the marks of squid tentacles and there is evidence to show that it may dive to depths of up to 600 fathoms (3600 ft.) in search of them. Within the head is a cavity containing a clear liquid oil called 'spermaceti', the function of which is not understood. Exposed to the air it quickly solidifies into a white wax once used for candle-making. A Sperm Whale is easily identified by its blow which occurs repeatedly while the whale is taking a breather near the surface. The spume is ejected through a single nostril which, being asymmetrically placed, causes the blow to emerge at a characteristic angle. A further characteristic of Sperm Whales is that they do not sink when dead, an advantage which endeared them to the old-time whalers with their open boats and hand harpoons; baleen whales, by contrast, have 'negative buoyancy' and must be pumped up with air before being left to float.

Dolphins

The remaining toothed whales found in Antarctic waters are not commercially important. They include the bottle-nosed, or beaked, whales (family Ziphiidae) and the dolphins (family Delphinidae). Most prominent among, and largest of, the dolphins is the notorious Killer Whale; these whales may be up to 30 ft. in length and are distinguished by a tall dorsal fin, which cleaves the water like a hatchet, by a vivid black and white colouring and by the formidable teeth in both jaws. The Killer is a notorious predator taking seals, penguins and other dolphins. Killers hunt, like wolves, in packs and are believed to attack the larger Baleen Whales for the sake of the tongue, their favourite delicacy. Like other dolphins the Killer has a large brain and appears to be highly intelligent. It is equipped with an echo-sounding system for navigation and can communicate with other Killers over a wide range of sonic and ultra-sonic frequencies.

Whaling industry

The whale has been hunted by man for food from time immemorial; as a regular industry whaling probably dates from the twelfth century when it was practised by the Basques in the Bay of Biscay. Today, the products of the industry add an important, though declining, contribution to the world's supply of food, the most important item being the oil of the Baleen Whale; much of this can be converted into edible fats, such as margarine, by a hardening process known as 'hydrogenation'. The oil of Sperm Whales, which is chemically different from that of the Baleen, is used for various industrial purposes; whale meat provides food for both humans and animals. Only the Japanese use it for human consumption in large quantities as they lack adequate meat protein from other sources. Nations such as the British have rejected it because of the faintly fishy flavour of its oil but consume it, perhaps unwittingly, as the basis of some made-up foods. Its by-products are used by the pharmaceutical industry for vitamin

The old whaling days – the hazards of the chase as depicted in William Scoresby's *The whale-man's adventures in the Southern Ocean*, 1850

and other extracts. Even the bones, rich in oil, can be pressure-cooked and the resulting meal used as a fertilizer or the basis of cattle food. The staples of the old-time whaling industry were oil and whalebone; the former lit the city streets of early nineteenth-century London and Boston, the latter was used for corsets, umbrellas, and numerous other manufactured articles. Spermaceti was used for making fine candles and ambergris, a product of digestion to be found occasionally in a Sperm Whale's intestine, was, and still is, much in demand by makers of cosmetics.

Whaling in Antarctic waters is of recent origin. It dates from 1904 when the first shore station was established at South Georgia to exploit the then abundant stocks of whales reported by various exploring expeditions. The industry developed rapidly and spread to the South Shetland and South Orkney Islands where processing was done aboard factory ships anchored as land stations. It soon became clear that measures would need to be taken to regulate the industry if the stocks of whales were to be conserved. In 1923 the British Colonial Office set up the Discovery Committee to undertake the necessary biological research. With the exception of the war years, 1939–45, this research on whales and the whole economy of oceanic life in the Southern Ocean has been carried on, the results being published in a series of high academic standing, the *Discovery Reports*. A new phase in Antarctic whaling opened in 1925 when whaling from factory ships capable of operating on the high seas for an entire season, unlicensed and independent of a shore base, began. By 1930 nearly the entire industry was

ANTARCTIC PELAGIC WHALING						Land Stations South Georgia
Year	No. of floating factories	No. of catchers	No. of Humpbacks taken	No. of Blue Whale units	Oil production in barrels*	Oil production in barrels*
1956/57	20	225	679	14,745	2,098,854	148,068
1957/58	20	237	396	14,850	2,146,206	171,432
1958/59	20	235	2,394	15,300	2,050,241	102,418
1959/60	20	220	1,338	15,512	2,050,892	97,546
1960/61	21	252	718	16,433	2,123,157	109,727
1961/62	21	261	309	15,253	2,001,961	49,815
1962/63	17	201	270	11,306	1,495,779	—
1963/64	16	190	2	8,429	1,299,476	41,282
1964/65	15	172	—	6,987	1,017,611	45,805
1965/66	10	128	1	4,085	634,299	9,964
1966/67	9	120	—	3,511	600,130	No whaling

* Barrel = 170 kg.

WHALING STATISTICS 1956/7 to 1966/7

pelagic with 41 factory ships and 200 catchers at work. Again fears were aroused as to whether the whale stock could survive such an onslaught. In 1931 an International Convention for the Regulation of Whaling was signed at Geneva and from then onwards restrictions have been in force. The war years gave the whales a temporary respite and a chance to build up numbers, but in recent years the scale of operations has increased again and a reluctance on the part of some participating nations to fix realistic quotas may well result in the total eclipse of the Antarctic's only staple industry.

Of the several nations formerly participating in the Antarctic whaling

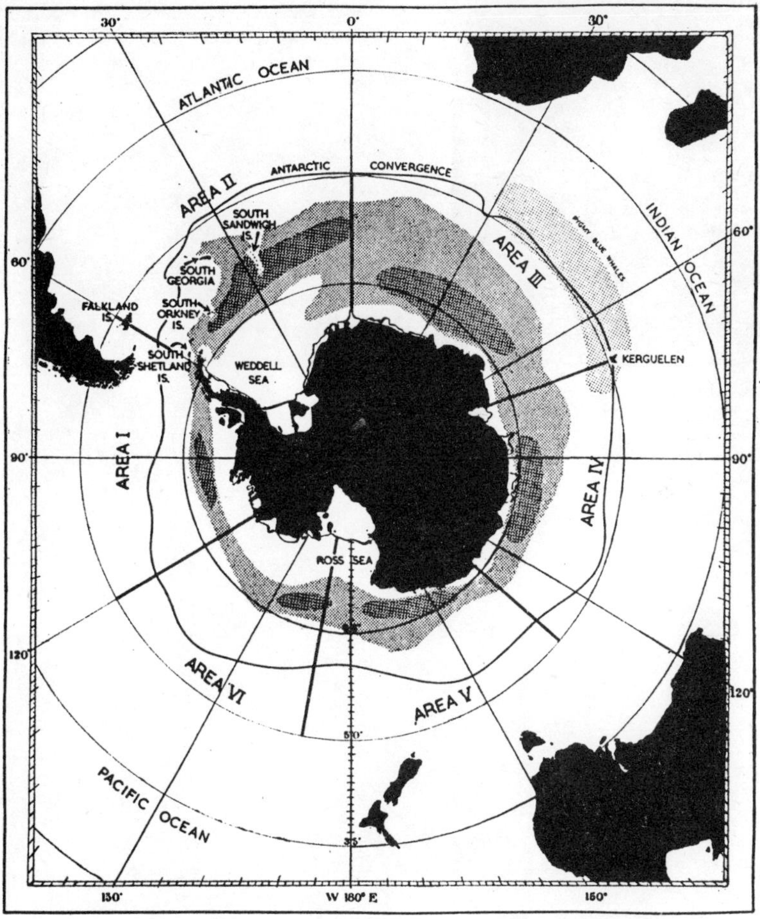

The Antarctic whaling grounds. The catches of pelagic whaling expeditions in the past forty years have mostly been in the continuous circumpolar zone (*shaded*)

industry, among which Britain was one of the most prominent, only two are now actively engaged, namely Japan and the U.S.S.R., sharing between them the International Whaling Commission's quota of 2469 Blue Whale Units.*

A modern pelagic whaling fleet is a formidable armada operating with the precision of a naval squadron in war-time. The centre of the fleet is the factory ship; her high main deck provides ample room for flensing and cutting up whales while below decks are numerous tanks for storing whale oil, meat meal, diesel oil for the ship's engines and even sea water for

* A Blue Whale Unit is calculated on the basis that 1 Blue Whale Unit equals 2 Fin Whales or 6 Sei Whales. This quota arrangement was fixed for the 1967–8 Antarctic season at the nineteenth meeting of the International Whaling Commission, 1967.

The former land-based whaling station at Leith Harbour, South Georgia. Antarctic whaling is now entirely carried out from floating factory ships

Cutting up whale blubber

Soviet whale factory ship *Sovetskaya Rossiya* with attendant catchers

ballast. To the aft of the vessel is a slipway up which the bodies of the whales are hauled for processing. Also below decks are boilers for cooking blubber, centrifugal separators for removing oil and residual glue water, machines for drying and bagging whale meal, refrigerating plant, laboratories, office and hospital accommodation, etc. One of the largest of modern factory ships, the Soviet Union's *Sovetskaya Rossiya*, is registered as 32,000 tons and is attended by 17 whale-catchers. The whale-catcher is small by comparison, perhaps only 500 tons or so; designed expressly to chase and overhaul the fast-swimming whales through the heavy seas of the 'filthy fifties' and 'screaming sixties', their diesel or diesel-electric engines can give over 3000 h.p. and attain speeds of up to 18 knots. From a distance the catchers are easily identified by the crow's nest, near the mast top, and the gun platform high in the bows, from which vantage-points the lookout and the gunner can spot, chase and eventually attack a surfacing whale. The ultimate weapon is the harpoon gun; invented by the Norwegian whaling magnate, Svend Foyn, in 1860, it came into general use in 1880 when it revolutionized the whaling industry, until then a hazardous business conducted from open boats with harpoons thrown by hand. The present-day harpoon has a grenade screwed to it to which is attached a time-fuse. This allows the harpoon to penetrate deep into the whale before exploding causing, so the experts claim, instantaneous death within seconds. Attached to the harpoon is a long line of manilla or nylon rope controlled from a winch below decks with which the whale can be played and eventually hauled alongside the catcher. A spring-loaded barb on the end of the harpoon prevents accidental withdrawal during the struggle and the consequent loss of the whale. Compressed air is now pumped into the body through the air spear, the number of the catcher scratched on the tail flukes, a metal foil reflector fitted to aid radar location of the whale and the carcass left to be picked up at a convenient time, freeing the catcher for the next chase.

In overall charge of the whaling expedition is the captain of the factory ship; his job is difficult and highly responsible; he must deploy his catchers to the best advantage over a wide area and take innumerable and complex decisions affecting the welfare of his men and the profitability of the voyage.

9
SEALS AND SEALING

The waters of the Southern Ocean are shared by the whales with only one other group of mammals – the seals. Known collectively to zoologists as the Pinnipedia (meaning animals with paddle-like feet), seals have a world-wide distribution and are to be found not only in both north and south polar regions but in temperate and tropical waters also, even in land-locked waters like the Caspian Sea and Lake Baikal. The word 'seal', like 'whale', covers numerous different species, 31 in fact, belonging to 3 main groups. Firstly there are the Eared Seals, or Otariidae, which include the Sea Lions and Fur Seals; secondly the Walruses, or Odobenidae; and thirdly the True Seals, or Phocidae. Walruses are not native to the Southern Hemisphere and we need not discuss them here. Eared Seals differ from True Seals not because of the visible external ears of the former but in the method of locomotion. All seals are well adapted to life in water but the Eared Seals least so; though powerful swimmers they are most active on land using their fore- and hind-flippers to gallop like normal quadrupeds. The True Seals, in contrast, are badly adapted for movement on land; the fore-limbs are too small to raise the animal off the ground and progress is made by a series of hitching caterpillar-like movements. The hind-flippers of the True Seal are stretched permanently backward making a splendid rudder in water but useless for land locomotion. These seals swim literally like fish with a powerful undulation of the rear part of the body. The four most typical Antarctic seals are all True Seals. Their distribution in the Southern Ocean is circumpolar but each group tends to have its own hunting-grounds and there is no competition between them.

True Seals

The most widely ranging of the Antarctic True Seals is the Leopard Seal. A solitary animal, normally inhabiting the outer fringes of the pack ice, the Leopard Seal often winters on the sub-Antarctic islands and is even seen occasionally on the beaches of south Australia and New Zealand. The long slim body (it may attain 12 ft. in length) is slate-grey in colour and spotted with yellow, and the large sharp teeth, which arm the reptilian head, account for the colloquial name. A predilection for young Adélie Penguins, which it shakes from their skins and swallows whole, has earned the Leopard Seal a bad reputation among the more sentimental of Antarctic travellers. Curiously enough, when out of the water, Leopard Seals and Adélies live side-by-side on the ice-shelves seemingly in perfect harmony.

The most abundant of Antarctic seals is the Crabeater; unlike the Leopard Seal it is gregarious, inhabiting the drifting pack ice and migrating to its northern edge in winter. Only occasional stragglers are reported further north than the pack-ice edge. Adult Crabeaters average about 8½ ft. long and weigh about 500 lbs. The brownish-grey back of the young Crabeater, with its chocolate-brown marking, fades gradually through the year to a creamy white. Little is known about the breeding habits of this seal; the pups are born in spring sometime between the middle of September and the beginning of November. 'Krilleater' would have been a more appropriate name for this species since the small shrimp-like krill form their staple diet. The interlocking cusps of the upper and lower teeth form a strainer which functions like the baleen in the mouth of the Blue Whale. Mummified bodies of Crabeaters have been found on the Antarctic continent high up in the dry valleys. Their age has been estimated at over 2000 years old. These were very likely young seals who failed to get away before the sea ice froze over in the autumn and were left to wander aimlessly, eventually dying of starvation.

Inhabiting the heavy pack ice round the edge of the Antarctic continent is the Ross Seal, named after the nineteenth-century polar explorer, Sir James Clark Ross. It is the rarest and least known of the Antarctic seals but can easily be recognized by its external appearance which is quite striking with its plump, shapeless body, large flippers, short head and large protruding eyes. Its chief food is probably squids. The curious cooing noises it makes have earned it the name of the 'Singing Seal'.

The most southerly seal of all is the Weddell Seal. It is usually found all round the Antarctic continent, sometimes on the fast ice within sight of land but normally on the pack ice. There are colonies both in South Georgia and the South Orkney Islands, and Weddell Seals are from time to time seen on other sub-Antarctic islands. The Weddell does not migrate, remaining in high latitudes right through the winter. Edward Wilson, zoologist on both Scott's expeditions, likened it in this respect to the Emperor Penguin, while the Crabeater he compared to the Adélie, both of whom winter out on the pack ice. Weddell Seals spend most of the winter months under the ice where the water temperature is considerably higher than that of the atmosphere. Their calls can be heard from sub-ice chambers which they ventilate with breathing-holes cut through the hard ice with canine teeth used in the manner of a circular saw. Weddell Seals

Weddell Seal

Leopard Seal

TRUE SEALS (Phocidae)

Common name	Specific name	Distribution
Southern Elephant Seal	*Mirounga leonina*	Sub-Antarctic and Antarctic islands north of the pack ice; very occasionally in the pack ice
Weddell Seal	*Leptonychotes weddelli*	Inshore waters of Antarctica and adjacent islands
Leopard Seal	*Hydrurga leptonyx*	Fringe of the pack ice and anywhere in the Antarctic between latitudes 50° and 80° S.
Ross Seal	*Ommatophoca rossi*	Pack ice
Crabeater Seal	*Lobodon carcinophagus*	Drifting pack ice south of the Antarctic Convergence

EARED SEALS (Otariidae)

Southern Fur Seal	*Arctocephalus australis*	Falkland Islands, southern Brazil to southern Peru and the Galapagos Islands
New Zealand Fur Seal	*Arctocephalus forsteri*	New Zealand and the sub-Antarctic Islands including Macquarie Island
Kerguelen Fur Seal	*Arctocephalus tropicalis gazella*	Islands north and south of the Antarctic Convergence from Ile St. Paul and Ile Amsterdam to the South Shetland Islands
South American Sea Lion	*Otaria byronici*	South America and the Falkland Islands
Hooker's Sea Lion	*Phocarctos hookeri*	Macquarie Island and Auckland Islands

ANTARCTIC SEALS

have black backs splashed with white, the stomach being grey streaked with white. Pups, born in September and October, are advanced in development being nearly 5 ft. long and weighing up to 60 lbs. Within two or three weeks they enter the water learning to eat crustaceans, the chief food of the species.

Breeding on many of the sub-Antarctic islands is the largest member of the group of True Seals, and indeed of all the seals, the Southern Elephant Seal. An adult male can reach 16 to 18 ft. in length and weigh between 2 and 3 tons. The cow is about half this size. Males are dark grey in colour fading during the year to a greyish brown; females are a darker brown. Quite the most outstanding physical feature of the male Elephant Seal (and the reason for its name) is its inflatable proboscis. Really an enlargement of the nasal cavity, it normally hangs downward over the mouth but can be erected to form a kind of cushion. This can also act as a sounding chamber amplifying the bull's roar to terrifying dimensions. Like the other seals we have described the Elephant Seal is designed for a life at sea. But at seasonal intervals these seals prefer to congregate on sub-Antarctic beaches to breed and to moult. The breeding season starts in early September. The bulls are the first to come ashore followed by the pregnant cows. The cows then gather into large groups called 'harems' numbering up to a thousand individuals. Each harem is ruled by a solitary bull who has ascendancy over his rivals in battle. Rulers of harems are known as beachmasters. In the larger harems the beachmaster may be attended by one or two assistant beachmasters. Battles between two contestants, usually occasioned by an invasion of territory, are often bloody but rarely fatal. After a preliminary challenge of tremendous trumpeting roars, the bulls, with probosces fully inflated, approach each other slowly, then rear up in a position of precarious balance. The object of the exercise is to throw an opponent off his balance and this is done by much biting and tearing. After 15 minutes or so one of the bulls will deflate his proboscis and retreat, pursued briefly by the victor. These battles continue throughout the season so that the beaches are not the best of places for rearing young pups. When a cow is about to give birth she wisely moves to the edge of the harem where her young are less likely to get trampled to death. Pups are weaned after about three weeks and are then deserted by their mothers. While they teach themselves to swim they have to live on their blubber. Moulting is an annual process which is also accomplished on the beaches where muddy pools, or wallows, among the tussock-grass provide a solace for itchy Elephant Seals.

Eared Seals

The Eared Seals, the other family of the order Pinnipedia found in the Antarctic, are represented by 3 species of Fur Seals and 2 species of Sea Lions. Of the Fur Seals only one breeds south of the Antarctic Convergence,

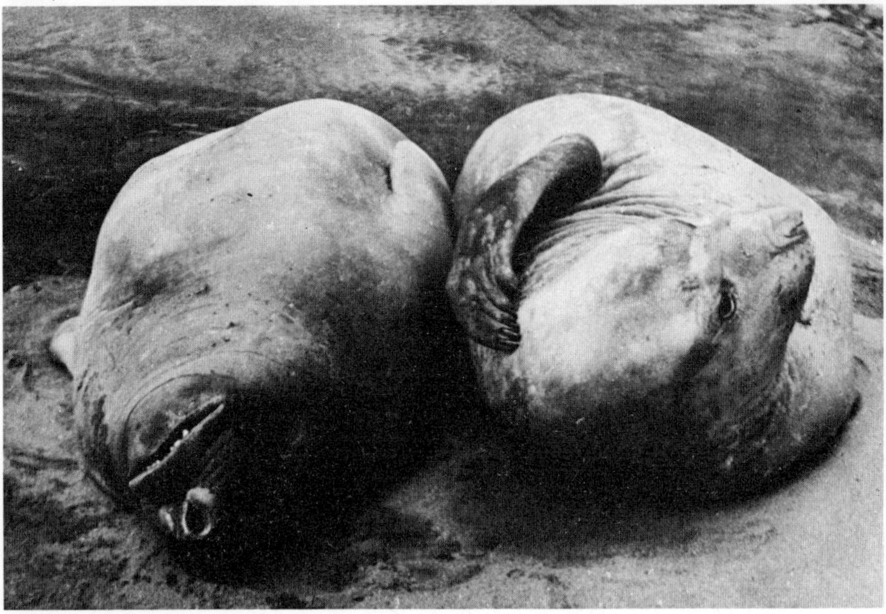

Elephant Seals

Herd of Elephant Seals,
Undine Harbour, Graham Land

Bull Elephant Seal roaring

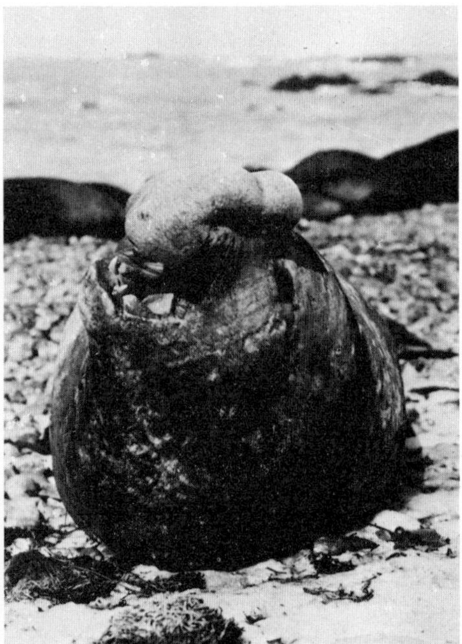

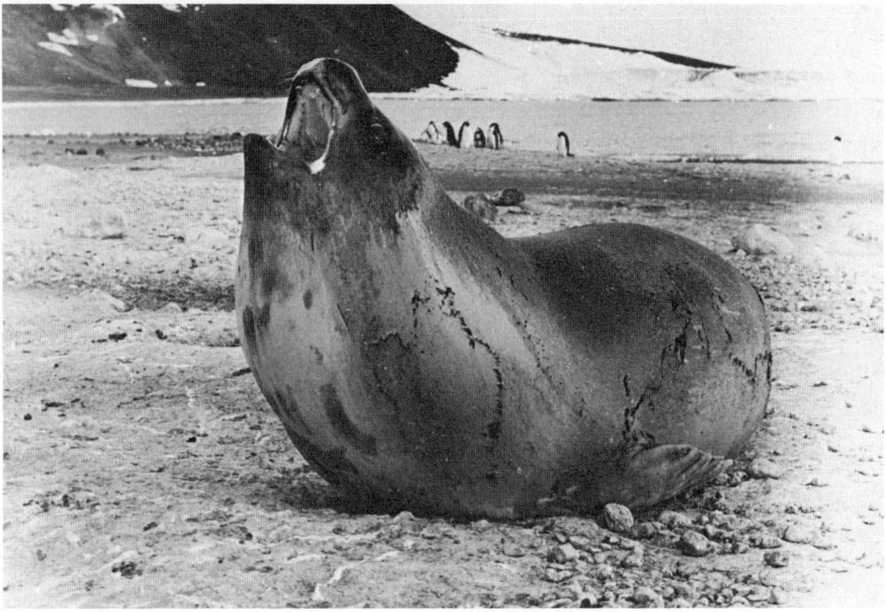

Ross Seal, rarest of the Antarctic seals

Fur Seals in tussock grass

the Kerguelen Fur Seal. All the members of the Fur Seal genus are rather similar in appearance and habits. Like the Elephant Seals, they are organized into beach harems. By comparison these are small with perhaps only 9 cows to 1 bull; competition between bulls is largely for the favourite territories by the water's edge. Their staple food is krill.

Fur Seals found in the sub-Antarctic are the Southern Fur Seal and the New Zealand Fur Seal, the former on the Falkland Islands, the latter on the sub-Antarctic islands of New Zealand.

Sea Lions are not found south of the Antarctic Convergence. The most common of the two sub-Antarctic species is the South American Sea Lion, plentiful in the Falkland Islands. Hooker's Sea Lion is found on the Auckland Islands, on Campbell Island and, occasionally, as a visitor to Macquarie Island.

The sealing industry

Like whales, seals have served the economic needs of man since time immemorial. To some Eskimo peoples the seal is still today the very staple of existence, its fur pelt providing warm clothing and its blubber food, its oil burned as fuel and its bones fashioned into tools or simply used as media of artistic expression. In the Antarctic there are, of course, no native peoples; such truly Antarctic seals as the Crabeater and the Weddell, the Ross and the Leopard Seals are largely protected from man the predator by the impenetrability of the pack ice. There are, however, two Antarctic seals of commercial value; one is the Fur Seal, a member of the family of Eared Seals, the other a True Seal, the Elephant Seal. The value of the Fur Seal lies in its skin; this can be used simply for the skin itself with the hair removed, and may be manufactured into leather coats and various small articles. What many people refer to as 'seal skin' is usually the soft undercoat of the seal, a fine velvety fur which remains when the long stiff outer, or guard, hairs have been removed. The value of the Elephant Seal lies in its great oil potential; the blubber from a member of this species may give up to two barrels, or one-third of a ton, of oil. Antarctic sealing today is non-existent, both the Elephant Seal and the Fur Seal being protected from commercial exploitation. The seal furs that grace the fashion houses of London, Paris and New York will most likely originate from the remote Pribilof Islands in the North Pacific Ocean, or the Atlantic coast of Canada. This has not always been so; there was a time when the islands of the Antarctic and sub-Antarctic were among the richest breeding-grounds of Fur Seals and Elephant Seals in the world. A history of wasteful exploitation, prompted by the crudest commercial greed, reduced the numbers of both species to a point of near extinction.

When sealing first began in the Southern Hemisphere is unknown; sealing voyages certainly became a regular occurrence in the early seventeenth century. Sealers are known to have visited the Falkland Islands

'Seal hunting on the pack-ice.' From Sir J. C. Ross's *Voyage . . . in the southern and Antarctic regions. . .*, 1847

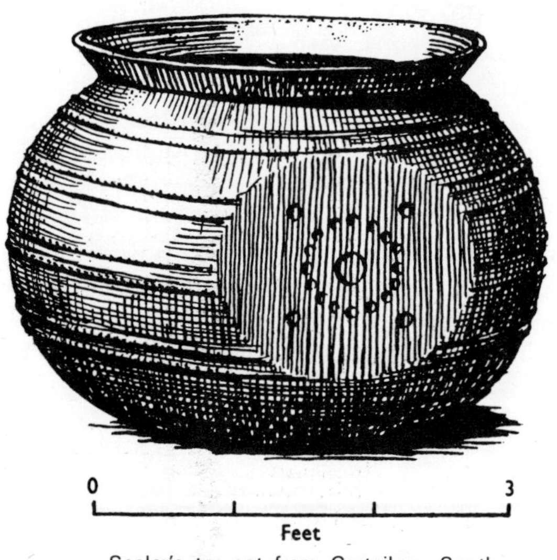

Sealer's try pot from Grytviken, South Georgia, now at the Scott Polar Research Institute

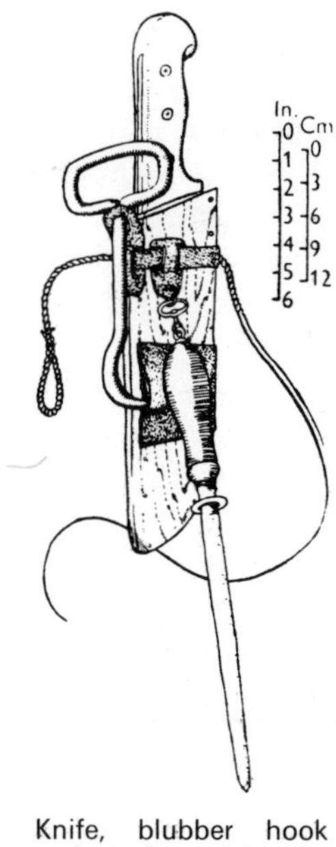

Knife, blubber hook and steel used by the sealers

as early as 1766 and it is very likely that the abundant Fur Seal populations of the islands off the west coast of South America were also being harvested about this time. The most profitable market for seal skins was then Canton, where, about 1750, an enterprising Chinese merchant discovered a technique for removing the coarse outer hairs of the skin leaving only the fine undercoat. As a result the demand for these skins rose enormously; the wealthier Chinese not only wore them on their backs but used them to adorn the walls and floors of their houses and insulate them from the cold. For some years this profitable market was largely in the hands of the Russians, the seal population of the Aleutian Islands, then a part of the Russian Empire, being decimated in the process. Soon the hunt was pursued into southern waters by both British and American vessels. From the Rio de la Plata to Cape Horn, from the Islands of Masafuera and Juan Fernandes off the Pacific coast of South America, to Archipel de Kerguelen and Iles Crozet and the Falkland Islands in the sub-Antarctic, the once teeming populations of Fur Seals were virtually obliterated. Sealing in the Antarctic proper dates from Captain James Cook's discovery of the 'Island of Georgia', or South Georgia as it is known today. Johann Reinhold Forster, the naturalist who accompanied Cook, described the great number of Elephant Seals and Fur Seals to be found on the island and even prophesied their possible exploitation. He certainly never foresaw the 'fur rush' that followed. Numerous British and American merchants were encouraged to fit out ships for the Fur Seal and 'elephant oil' trade. By the end of the eighteenth century both British and American sealers were taking enormous hauls from South Georgia alone. One American sealer, Edmund Fanning, in the *Aspasia* of New York, secured, in 1800, a record catch of 57,000 Fur Seals. Captain James Weddell, a British sealer, writing in 1825, estimated the total number of skins taken from South Georgia as not less than 1,200,000 and the quantity of Elephant Seal oil at 20,000 tons.

The methods used by the early sealers seem brutal to us, but this was an age less sensitive to animal suffering than our own. Another British sealer, Robert Fildes, a cultured man, not lacking in sensitivity, describes in his journal the problems involved in attaining these 'fur lined' beaches. 'It was impossible to haul up a boat without first killing your way, and it was useless to try to walk through them, if you had not a club in your hand to clear your way and then twas better to go two or three together to avoid being run over by them, tho there was no fear of them biting for they were quite harmless.' No attempt was made to cull the herd so as to leave an adequate breeding stock for the future; whole populations were wiped out. Old male seals, or 'wigs' as they were called by the sealers, the females, or 'clapmatches', and the young pups were all indiscriminately lanced or clubbed to death and flayed on the spot. An expert sealer, it was said, could skin 60 seals in an hour. The final refuge of the Antarctic seals, the South Shetland Islands, was discovered in 1819 and the same sad sequence of events followed. By 1829 a British captain, W. H. B. Webster, could write in his journal 'The harvest of the seas has been so effectually reaped that not a single fur seal was seen by us during our visit to the South Shetland Group.' So ended, ingloriously, the golden age of Antarctic sealing. From now on Elephant sealing was to become the predominant activity in the waters of the sub-Antarctic. The techniques were similar, the creatures being driven to the water's edge and then killed with a couple of musketballs through the mouth. The blubber was then chopped up and boiled in large cauldrons, or try-pots, mounted on bricks. A few of these giant pots, which can hold up to 200 gallons, are still to be seen on South Georgia. Fuel to heat the pots consisted of penguin skins or blubber. The pots were arranged in series, the oil being run into a second pot after the first boiling, boiled once or twice more and the refined oil resulting run off into wooden casks and shipped home. By 1872 the Elephant Seal industry was virtually dead, the stocks so depleted that it was no longer profitable to hunt them.

By the turn of the present century the Elephant Seal population of South Georgia had begun to recover in numbers and the Falkland Islands Government, whose jurisdiction included this remote British outpost, introduced a Seal Fishery Ordinance in an effort to put the industry on a rational basis. Killing of seals was to be licensed and restricted to adult males; annual reports on the state of the rookeries were made compulsory. Today there is a restricted hunting season from early September to mid-November, this being the period when the seals come ashore to breed and produce the most oil; the annual catch is limited to 6000. Since 1910 about a quarter of a million seals have been harvested at South Georgia yielding 74,000 gallons of oil. Population statistics seem to indicate that the Elephant Seals are now holding their own and the prospects for their future seem good.

There are signs, too, that the Fur Seals are gradually re-establishing themselves not only in the area of South Georgia but on the South Orkney Islands and the South Sandwich Islands also. Providing a rational policy of exploitation is agreed on there seems every reason to hope that the stocks of this species also will be maintained. That the sealing industry of the Antarctic will ever again attain its former importance seems unlikely; the northern Fur Seal industry can well cope with the demand for furs. But to a world increasingly aware of the need to conserve its wildlife, this re-establishment of two species of seal which narrowly escaped extinction must be a source of satisfaction.

10

THE ISLANDS OF THE ANTARCTIC AND SUB-ANTARCTIC

Our account of the Antarctic regions so far has concentrated, in the main, on the Antarctic continent and on the surrounding Southern Ocean. We turn now to the islands of the Antarctic and the sub-Antarctic scattered irregularly over the vast expanses of ocean that girdle the Earth south of latitude 40°S. Few in number, insignificant in area and mostly uninhabited, these bleak, windswept, rocky outcrops seem, at first sight, to be of little real interest or use other than as meteorological stations or refuges for seals, penguins and other sea birds. But their true significance was appreciated more than a century ago by that pioneer of Antarctic botany, Sir Joseph Hooker. Writing of the sub-Antarctic islands of New Zealand in his *Flora Antarctica* (1844) he said: 'It will appear that islands so situated furnish the best materials for rigid comparison of the effect of geographical position and the various meteorological phaenomena on vegetation, and for acquiring a knowledge of the great laws according to which plants are distributed over the face of the globe'. Biological research today is still very much concerned with the effects of climate and situation on plant as well as on animal life. For in a world where man has the power to speed up the whole process of evolution and thereby influence, for better or for worse, the future not only of his own species but of all other living things on this planet, it is becoming increasingly urgent to discover the basic rules affecting environmental change. These rules can best be found by studying simple, unspoiled communities of plants and animals such as those which populate the islands of this region. Indeed, the sub-Antarctic and Antarctic islands have the one great advantage of being among the few remaining places on Earth where the native fauna and flora are still relatively free from interference by man.

We shall discuss briefly the effects of climate on the pattern of island life and landscape before going on to describe the individual island groups and their history. But firstly we need to restate our definition of what constitutes the islands of the Antarctic and sub-Antarctic.

Definitions

Climatically the islands group themselves into two main zones; those to the south of the Antarctic Convergence, corresponding approximately with the northerly limit of the pack ice in winter, we class as Antarctic islands. They include the following: South Georgia, the South Sandwich Islands, the South Orkney Islands, the South Shetland Islands, Bouvetøya, Heard Island, the Balleny Islands, Scott Island and Peter I Øy. To the north of, but verging on the convergence, are the Prince Edward Islands, Iles Crozet and Archipel de Kerguelen; the climate of these islands is sufficiently affected by the cold upwelling currents of the convergence to be regarded as Antarctic in character. All the remaining islands north of the convergence we shall describe as sub-Antarctic. They include: the Falkland Islands, Tristan da Cunha, Gough Island, Ile St. Paul, Ile Amsterdam, Macquarie Island and the sub-Antarctic islands of New Zealand including the Auckland Islands and Campbell Island. This sub-Antarctic zone has a boundary to the north called the 'Subtropical Convergence'. Not so well defined as the Antarctic Convergence, it corresponds approximately with the northernmost limit of iceberg drift.

The relationship between climate and life

Island climates south of the Antarctic Convergence have much in common with the climate of Antarctica – they are cold with low precipitation. The landscapes of Antarctic islands are typified by perpetual snow and glaciers. Island climates of the sub-Antarctic zone are much less severe and there is little or no permanent snow and ice. Here the open, ice-free ocean has an ameliorating effect; the temperature range is small, rainfall is more general and the air full of moisture. Cloud, fog and mist are nearly always in evidence and sunny days, for visitors, are red letter days. The sub-Antarctic is also one of the world's windiest regions, corresponding as it does with those latitudes of westerly winds which seamen know as the 'roaring forties'.

These climatic changes from sub-Antarctic to Antarctic are of course very gradual; each zone could be divided into a number of intermediate sub-zones. The effect of the climatic gradient on island life is best illustrated by voyaging southwards from the Falkland Islands along the string of islands which link South America with the Antarctic continent. The climate of the sub-Antarctic Falkland Islands is regarded by biologists as 'cold temperate', that is, for over half the year average temperatures exceed 5 °C. and rarely drop below freezing-point. Rainfall, too, is plentiful.

Month	Falkland Islands	South Georgia	Marion Island	Archipel de Kerguelen
July	2.2	−1.4	3.6	1.9
December	8.3	3.5	6.4	5.8
Average	5.7	1.6	5.4	4.3

Month	Heard Island	Macquarie Island	Auckland Islands	Campbell Island
July	−0.3	2.9	5.9	4.5
December	2.2	5.9	10.3	8.2
Average	1.1	4.5	8.2	6.7

MEAN MONTHLY TEMPERATURES AND YEARLY AVERAGES (°C) FOR SOME SUB-ANTARCTIC ISLANDS

Typical of this type of damp, cool climate are grassy steppes and peaty moorland with many flowering plants. Trees are lacking but mainly because of the strong prevailing westerly gales. The Falkland Islands also lack native land mammals because of their isolation from the mainland. But sea birds are abundant and penguins and Elephant Seals breed among the tall tussock-grass.

Lying in the same latitude as the Falkland Islands, but south of the Antarctic Convergence, is the island of South Georgia, glacier-covered and typically Antarctic in appearance. Tussock-grass and moorland plants are here confined to the lower slopes of the icy mountains which are typical of the island. Very few species of flowering plants are to be found, the largest form being a grass. However, in sheltered places, grass heaths develop and peat bogs exist in some valleys. Land animals are equally scarce – only 1 native bird and some 70 insect species. South Georgia has no month with a temperature above 5 °C., and for about half the year average temperatures are below freezing. These are not conditions favourable for temperate vegetation. Only round the coasts is there the usual proliferation of life – sea birds, penguins and seals.

As one approaches the Antarctic continent the severity of the climate, and with it the impoverishment of life, is accentuated. On the South Shetland Islands one finds only two species of higher plant, a grass and a cushion-plant growing in a few spots favoured by the Sun. Plant life is now entirely dominated by the mosses and lichens. Land life is reduced to a few primitive and flightless insects and other lowly creatures. In this zone of the Antarctic the mean temperature of the warmest month is only just above freezing-point and average monthly winter temperatures can fall to as low as −10 °C. Snow and ice cover the land and only sea-dependent animals can flourish.

This same climatic gradient of increasing life and decreasing cold can be observed elsewhere throughout the region. The climate of southern New Zealand, for example, is roughly comparable with that of southern South America; Campbell Island can be compared with the Falkland Islands, Archipel de Kerguelen with South Georgia. Everywhere throughout the sub-Antarctic and the Antarctic the variety of plant and animal life correlate in general with average temperatures, although, of course, the size, height and distance of islands from neighbouring continents will affect the situation to some extent.

Just as there is a pattern of impoverishment and dwarfing of the different forms of life from north to south, so there is a fairly regular latitudinal zoning of plant and animal species on the islands. The tussock-grasses, peatbogs and cushion-plant communities of the Falkland Islands will be found, though with varying dominant species, all over the sub-Antarctic islands eastward to Macquarie Island. Similarly, the seals, penguins and other marine birds, which are already described in an earlier chapter, each have their characteristic zones of distribution.

Problems of distribution

Remote islands have fascinated biologists ever since Charles Darwin published his studies of the animal life of the Galapagos Islands in the revolutionary *Origin of Species* (1859). Far removed from neighbouring continents, such islands support forms of life which, cut off from their parent stock, have evolved into peculiar species found nowhere else in the world. The islands of the sub-Antarctic provide many fascinating examples of these endemic plants and animals. Gough Island, for example, has 6 species of native land bird, several of which are represented by distinct sub-species on neighbouring islands. The pattern of distribution of plants and invertebrate animals is of especial interest. It would certainly appear that the islands have served, and are still serving, as biological stepping-stones linking continents as far apart as South America and Australia. For example, the cushion-plant bogs so typical of the Falkland Islands, the Auckland Islands and Campbell Island are akin to similar plant communities in Tierra del Fuego and New Zealand. The invertebrate animals of Gough Island nearly all have South American relatives. Some remarkable detective work by biogeographers has resulted in numerous theories as to how these various forms of land life succeeded in establishing themselves on such small, scattered islands in the first place. Many small seeds and spores are carried by the predominantly westerly winds from the direction of South America. Other seeds may be carried by birds or by the circumpolar ocean current known as the 'West Wind Drift'. Insects, too, are transported by westerly winds and seem to originate from South America. Whether these plants and insects are all newcomers since the last ice age, or whether any are relicts from before this geologically distant period is a matter for debate. There is fossil evidence from Archipel de Kerguelen to indicate that the islands, like adjacent Antarctica, supported a flourishing temperate forest vegetation millions of years ago, and it is possible that land connections, now submerged, then existed which would serve as biological causeways between Antarctica and the surrounding islands and continents.

Island geography

The second part of our chapter deals descriptively with the various island groups repeating, for the sake of convenience, some information provided elsewhere in this book. We start in the South Atlantic Ocean with the Falkland Islands and the islands of the Scotia Arc and then progress in an easterly direction by way of the islands of the Mid-Atlantic Ridge – Tristan da Cunha, Gough Island and Bouvetøya – to the Indian Ocean. Here we shall visit islands extending as far north as latitude 37–38° (Ile St. Paul and Ile Amsterdam) and as far south as 53° (Heard Island). Next, in the Pacific, we shall discuss a number of small islands adjacent to New Zealand as well as lonely Macquarie Island. In conclusion we shall very briefly note a few scattered off-shore islands of Antarctica which are of some scientific and historical interest – the Balleny Islands, Scott Island and Peter I Øy.

1. THE FALKLAND ISLANDS AND THE ISLANDS OF THE SCOTIA ARC

Linking the southern tip of South America, Tierra del Fuego, with the Antarctic Peninsula is a great, curving island arc. Believed to be geologically one with the South American Andes, these islands are the exposed peaks of submarine ranges and are known to geologists as the 'Scotia Arc'. They include Staten Island, the Burdwood Bank, Shag Rocks, South Georgia, the South Sandwich Islands, the South Orkney Islands and the South Shetland Islands. The Falkland Islands, which lie on the continental shelf of South America, are not geologically allied to the Scotia Arc and are believed to have affinities with the rock formations of South Africa. Nevertheless, this latter group of islands has many sub-Antarctic characteristics and is historically and politically connected with Antarctica. South Georgia and the South Sandwich Islands are in fact dependencies of the British colony of the Falkland Islands; the South Orkney Islands and the South Shetland Islands are part of a separate colony known as The British Antarctic Territory, which forms a wedge whose apex is the South Pole, whose sides are the meridians 20° and 80°W. longitude, and whose base is latitude 60°S. Both colonies are administered from the Falkland Islands.

The Falkland Islands (latitude 51°–53°S., longitude 57°–62°W.)

The Falkland Islands lie in the South Atlantic Ocean about 300 miles from the mainland of South America. They comprise 2 large islands, East Falkland and West Falkland and 200 smaller islands having a total area of 4700 square miles. The climate is not one of violent extremes – it is never very warm nor yet excessively cold. Rain falls on between 16 and 21 days every month and snow can fall at any time throughout the year, though total amounts are light and there is no permanent snow cover. Average annual rainfall is about 25 inches. South-westerly winds predominate and are frequent; it is these salt-laden winds which prevent tree growth on the islands.

The native vegetation is grassland which, with various species of heath and dwarf shrub, form moorlands similar to those found in the highlands of Britain. These grasslands are the basis of the present sheep-farming industry, the Falkland Islands' only natural wealth, for there are no mineral resources.

The wildlife of the Falklands Islands is an abundant one. Over 60 different species of birds nest here, many on the off-shore islands. Some, like Wilson's Petrel and the Sooty Shearwater are marine birds; land birds include two wrens and a robin. Both Antarctic and sub-Antarctic species of seals and penguins breed on the islands including the Magellan Penguin found nowhere else in the sub-Antarctic.

The human population is largely British in origin, fluctuating between 2000 and 2300 in number. More than half the inhabitants live in the capital, Stanley. This is the centre of the administration and here are the local offices of the British Antarctic Survey whose responsibility it is to refuel and equip southward-bound British Antarctic expeditions. Also at Stanley is the headquarters of the Falkland Islands Company which has important agricultural, commercial and shipping interests in the islands.

The Falkland Islands have had a chequered history since they were first sighted by the English captain John Davis in 1592. The French then occupied the islands in 1764 (they named them Iles Malouines after the

St. Malo fishermen who used to visit them). Later they were settled by both the Spanish and the British, the disputed sovereignty nearly leading to war. In 1832 the British reasserted their claim to the islands and have maintained it ever since. The original Spanish claim is still upheld by Argentina which regards 'Las Islas Malvinas' as part of her national territory.

South Georgia (latitude 54° 20′S., longitude 36° 40′W.)

Eight hundred miles south-east of the Falkland Islands lies South Georgia, Britain's oldest Antarctic possession. It is a long narrow island, 100 miles in length and 20 miles across, consisting of a chain of snow-covered mountains the highest of which is Mount Paget (9625 ft.). The intervening valleys are filled with glaciers which in some places stretch down to meet the sea. The climate is cold and damp with summer temperatures rarely exceeding 5.4 °C. Though conditions are nowhere as severe as those prevailing on the Antarctic continent the climate is generally unpleasant with mists and strong northerly winds predominating. As we have noted elsewhere life on this island is intermediate between that of the sub-Antarctic islands and that prevailing in the more northerly coastal regions of Antarctica. Tussock-grass characterizes the swampy ground near the sea and beyond this is a tundra-like region with various grasses, burnets and tiny flowering plants such as buttercups and bedstraw. Elsewhere mosses and lichens predominate. Coastal seaweed, or kelp, is very conspicuous. The only native land animals are some limited species of insect life but, as elsewhere in the Antarctic, coastal life is rich in birds and seals.

South Georgia has a recorded history reaching back to the sixteenth century when it was reputedly sighted by the navigator Amerigo Vespucci. It was next seen by an Englishman, Anthony de la Roché, in 1675. But

Species of tussock grass (*Poa flabellata*) dominant in coastal regions of the Falkland Islands, South Georgia, Gough Island and Tristan da Cunha

In 1965 the British Combined Services Expedition retraced Sir Ernest Shackleton's historic trek across the mountains of South Georgia. They also climbed Mount Paget, the island's highest peak. *Above:* the achievement of the summit; *below:* the view looking north from the summit

it was Captain James Cook, during his famous second voyage in search of an Antarctic continent, who first surveyed the island's inhospitable coastline in 1775 and claimed it for the British Crown. Cook was not much impressed by the value of these "lands doomed by Nature to perpetual frigidness' where 'not a tree was to be seen, nor a shrub even high enough to make a toothpick'. But his reports of rich stocks of seals and whales led to the establishment first of a sealing industry and then, when this declined, a thriving whaling industry. Today the stocks of whales are depleted and the island remains a sanctuary for the seals whose pelts and oil first led to its exploitation. Between 1951 and 1957 detailed surveying of the island was carried out by Duncan Carse. Memories of Sir Ernest Shackleton's dramatic trek across the island in 1916 to get help for his companions on Elephant Island were revived in 1965 when a British Combined Services Expedition, under the leadership of Lieutenant-Commander M. K. Burley, R.N., retraced the historic route.

South Sandwich Islands (latitudes 56° 18′–59° 28′S., longitudes 26° 14′–28° 11′W.)

This group, the second of the Falkland Islands' two dependencies, forms the eastern section of the Scotia Arc. It is the only volcanic island arc in the Antarctic. Several of the islands are volcanically active and as recently as 1962 a submarine eruption covered the sea for miles around with a carpet of light-coloured pumice. The sheer cliffs and cone-like profiles of these islands are very characteristic and, though largely covered by ice, there are considerable patches of ice-free rock. By and large they are a barren group displaying even less plant and animal life than the South Orkney Islands lying further still to the south. Only in the region of the warm and active fumaroles does life burgeon unexpectedly with green mats of mosses and liverworts. The grass, *Deschampsia antarctica*, has been found on Candlemas Island. Numerous sea birds breed on the cliffs and elsewhere; penguin rookeries are also numerous. Elephant, Weddell and Leopard Seals have been seen on the beaches.

It was Captain Cook who, in 1775, discovered the most southerly of these Islands – Vindication, Candlemas, Saunders, Montagu, Bristol and Southern Thule – and named them after Lord Sandwich, the then First Lord of the Admiralty. Mistakenly he judged them to be an extension of the Antarctic continent. The more northerly Islands, Zavodovski, Visokoi and Leskov, as their names suggest, were discovered in 1819 by a Russian expedition led by Captain Thaddeus von Bellingshausen. Collectively they are known as the Traversay Islands. Since then sealers, whalers and numerous scientific expeditions have been in the region but few landings on the islands have been made because of the heavy swell. In 1961–2 and 1962–3 successful surveys were made by the British Antarctic Survey using ship-based helicopters.

South Orkney Islands (latitude 60° 30′S., longitudes 44° 25′–46° 10′W.)

Lying 454 miles southwest of South Georgia and 293 miles north-east of the northern tip of the Antarctic Peninsula, this rugged mountainous group consists of two major islands, Coronation Island and Laurie Island, two minor islands, Powell and Signy Islands, and a number of scattered off-lying islands. Rising precipitously from the sea to irregular ridges and sharp peaks the islands have a barren and impenetrable appearance. The larger islands are characterized by numerous glaciers which in places flow down to sea-level. Only Signy Island, possibly because of its sheltered position, has but little ice. The South Orkneys are typically Antarctic with a harsher climate than the more southerly South Shetland group. The climate can vary considerably from season to season depending on the proximity of the Weddell Sea pack ice. The flora and fauna reflect the climate. Olive-green lichens cover exposed areas of rock and green algae flourish near the penguin rookeries. Elephant and Leopard Seals are plentiful and there is the usual rich bird population. Among the penguins the Adélie is the most abundant.

The South Orkney Islands were jointly discovered in 1821 by two sealers, an Englishman, George Powell, and Nathaniel Palmer, an American. The first scientific work on the group was carried out by W. S. Bruce's Scottish National Antarctic Expedition in 1903–4 when a meteorological station was established on Laurie Island. This station was subsequently handed over to the Argentine Meteorological Service which has maintained it ever since (station 'Orcadas'). On Signy Island there is a base of the British Antarctic Survey.

Two of the barren South Orkney Islands, linked by floating ice: Signy Island viewed from Coronation Island

South Shetland Islands (latitudes 61°–63° 30′S., longitudes 53° 30′–62° 45′W.)

This, the most southerly group of the Scotia Arc, forms a north-east trending chain separated from the Antarctic Peninsula by the deep Bransfield Strait. It includes Elephant and Clarence Islands, King George Island, Deception Island and Livingston Island, as well as innumerable islets and rocks. Like the South Orkneys, the South Shetlands are administratively a part of The British Antarctic Territory. They are characteristically mountainous and ice-covered providing some of the finest scenery in the Antarctic. The most unusual member of the group is Deception Island, at one time a favourite rendezvous for sealers. Almost round in shape and about 9 miles in diameter, it rises to an average height of 1000 ft. A crater-like basin of volcanic origin forms the centre. Its beaches are famous for their sulphur springs – hot enough to boil an egg, it is claimed. This relatively ice-free island forms a perfect harbour for shipping, the wall of the cone being breached by a narrow entrance known as 'Neptune's Bellows'. In November 1967 a series of volcanic eruptions sent up columns of black smoke and ashes high into the air and culminated in the speedy evacuation of the Argentine, Chilean and British bases on the island.

The climate of the South Shetland Islands is relatively mild for the Antarctic, though lichens and mosses are still the predominant form of plant life. Shore life is akin to that of the South Orkneys.

The islands were first sighted by William Smith, a British sealer, in 1819. On a return voyage he took possession of King George Island in the name of George IV. In 1820 Smith returned again, this time under the

Aerial view of Deception Island, South Shetland Islands. The narrow entrance, 'Neptune's Bellows', is seen in the middle foreground

The volcanic eruption at Deception Island in November 1967, which caused the speedy evacuation of the scientific stations

orders of a British naval captain, Edward Bransfield, who was the first person to examine and chart a part of the adjacent Antarctic mainland.

2. ISLANDS OF THE MID-ATLANTIC RIDGE

Unlike the Falkland Islands and the islands of the Scotia Arc, which are closely linked with the continents of Antarctica and South America, our next group of islands is what geographers would describe as 'oceanic', that is, they are far removed from the nearest continental mainland. Such are the islands of the Mid-Atlantic Ridge, one of the largest submarine mountain ranges in the world, stretching from Greenland and Iceland in the Northern Hemisphere to within a thousand miles of Antarctica itself. Tristan da Cunha and its neighbouring islands lie in the north of this sub-Antarctic zone; only Bouvetøya, lying south of the Antarctic Convergence and close to the pack ice, is typically Antarctic. Peaks of the submarine ridge, these islands are all volcanic in origin, Tristan da Cunha and Bouvetøya rising from the ocean as characteristic cones.

Tristan da Cunha and Gough Island (latitudes 37° 05′–40°S., longitudes 15° 20′–10°S.)

Tristan da Cunha, with its neighbouring islands of Inaccessible and Nightingale, and Gough Island 230 miles to the south-east, lie in the South Antlantic Ocean almost midway between South Africa and South America. The only permanent population is on Tristan da Cunha, the other islands being uninhabited. There is a meteorological station on

Gough Island maintained by the South African Weather Bureau. Politically the Tristan da Cunha group of islands is part of the British colony of St. Helena. With their warm wet climate the group scarcely qualifies to be classed as sub-Antarctic. Gough Island, for example, boasts a luxuriant vegetation from shoreline almost to the top of its 3000 ft. summit. But like most other oceanic islands there is the characteristic poverty of species everywhere – no freshwater fish, no amphibians, no reptiles nor land mammals. But unlike many other islands these have largely escaped the destruction of their native plants and animals by man. And it is still possible to compare different species of an animal as between one island and another. For example, different species of a flightless rail (a bird resembling an English moorhen) inhabit Inaccessible and Gough. The animal and plant life of the group is of particular interest to science since it is still sufficiently distinctive for its continental origins to be traced. As well as native land plants and animals there are breeding colonies of the near-extinct Kerguelen Fur Seal as well as Elephant Seals. As always, sea birds form the most numerous species of animal life and breed in vast numbers. The population of Greater Shearwaters on Nightingale Island is said to be so vast that the birds are compelled to lay their eggs in the open instead of in their normal burrows. The pairs of Rockhopper Penguins breeding on Gough Island have been estimated at over 2 million.

The Tristan group was discovered by the Portuguese early in the sixteenth century, the first known landings dating from 1643. During the eighteenth and nineteenth centuries sealers were very active in the group and there must have been innumerable unrecorded landings. Tristan da Cunha itself has been inhabited almost continuously since 1810. It took a series of volcanic eruptions in 1961 to force the tenacious population to take temporary refuge in Great Britain; some have since returned and make a living from catching lobster and crawfish. Few scientists have spent much time among the islands. In 1955 the Gough Island Scientific Survey, an expedition from Cambridge University, spent six months on this island mapping it and studying the wildlife.

Bouvetøya (latitude 54° 26′S., longitude 3° 24′E.)

This, the southernmost island of the Mid-Atlantic Ridge, consists of a single volcanic cone, 3068 ft. high and 19⅓ square miles in area. An all-enveloping ice-cap reaches sea-level at places, and thick cloud normally obscures it from view. Temperatures rarely exceed 2 °C. in summer and average about −1.5 °C. in winter. Snowfalls are frequent. The flora is typically Antarctic – mosses and lichens – despite its relatively high latitude. There are colonies of Elephant Seals and Fur Seals and Macaroni and Ringed Penguins breed here. The Adélie, normally a more southerly breeding penguin, is also reported.

Bouvetøya's north-western extremity, Kapp Circoncision, was discovered in 1739 by a French navigator, Bouvet de Lozier, who believed it to be part of a supposed southern continent. A number of expeditions sighted the island subsequently but it was not until 1927 that a Norwegian scientific party was able to land and make a serious study of the island. It was subsequently annexed to Norway. Attempts by the government of South Africa to establish a meteorological station there have so far proved unsuccessful.

3. ISLANDS OF THE INDIAN OCEAN SECTOR OF THE SOUTHERN OCEAN

In the Indian Ocean sector of the Southern Ocean there lie a number of scattered islands and archipelagos, all volcanic in origin. Rising steeply from the submarine Atlantic-Indian and Crozet Ridges are Marion and Prince Edward Islands and their neighbours, Iles Crozet, 680 miles due east. Midway between South Africa and Australia lies Archipel de Kerguelen, 1000 miles east-south-east of Iles Crozet. To the south of the Kerguelen group lies Heard Island and to the north lie Ile St. Paul and Ile Amsterdam.

Ile St. Paul and Ile Amsterdam (latitudes 38° 43′–37° 52′S., longitude 77° 30′E.)

Like Tristan da Cunha these two volcanic islands have a temperate climate with an average summer temperature of 16 °C. and an average winter temperature of between 11 and 12 °C. Rainfall is moderate with only a sprinkling of snow in winter. Considerable cloud cover and strong winds are typical in winter. The flora is relatively abundant (though mostly introduced), the man-high tussock-grass (*Spartina arundinacea*) shared with Tristan, being typical. Unusual for these regions is the scattering of wind-stunted trees. The fauna includes insects and swarms of flies; in addition man has introduced rats, mice and cats which now roam the islands in the wild state. Both islands have huge rookeries of Rockhopper Penguins and there are Elephant Seals and Fur Seals. The islands have a commercially important crawfish industry.

A French possession since 1893, the islands have been known to seamen since the sixteenth century and being on the route between Europe and the East Indies were frequently visited.

Marion Island and Prince Edward Island (latitudes 46° 49′–46° 59′S., longitudes 37° 35′–37° 55′E.)

Lying 900 miles to the south-east of Capetown these islands, collectively known as the Prince Edward Islands, are the twin peaks of a submerged volcano. Marion is the larger of the two, with an area of about 84 square miles and rising to a height of 3890 ft. in Jan Smuts Peak. Prince Edward Island, to the north, is about half this size. The climate of the region is characteristically stormy and cloudy with frequent rain, and there is permanent snow on Marion above 2000 ft. The vegetation is akin to that on neighbouring Iles Crozet and Archipel de Kerguelen with a species of tussock-grass called *Poa Cookii* and a boggy moss vegetation on the hillsides as well as the edible Kerguelen Cabbage. Elephant Seals are plentiful and Leopard Seals visit the islands. Four species of penguin are found (Rock-hopper, King, Gentoo and Macaroni). Various species of albatross and petrel abound, one of which is native to the group (*Chionis minor marionensis*).

The islands were discovered in 1772 by a Frenchman, Captain Marion-Dufresne, but were actually named by Captain Cook in 1776. They are, today, under the sovereignty of the Republic of South Africa which maintains a meteorological station on Marion Island.

Iles Crozet (latitudes 46°–46° 30′S., longitudes 50° 30′–52° 30′E.)

This archipelago of volcanic islands lies approximately half-way between the Prince Edward group to the west and Archipel de Kerguelen to the east. It is composed of two groups of islands about 60 miles apart. The eastern group consists of two islands, Ile de la Possession and Ile de l'Est; to the west lie one main island, Ile aux Cochons, and two small islets, Les Apôtres and Iles des Pingouins. The main islands are all mountainous, the peaks of Ile de l'Est rising to 6500 ft. The surrounding waters are made dangerous by submarine reefs bristling with rocky needles. The climate is slightly warmer than that of Archipel de Kerguelen and there are virtually no glaciers. The incessant westerly winds are the most characteristic feature of the climate. The plants and animals are similar to those of Archipel de Kerguelen with tussock-grasses at shore-level, cushion-plants at higher levels, interspersed with the Kerguelen Cabbage and abundant lichens and mosses. The usual marine life typifies the coastal regions; Elephant Seals breed here and Leopard Seals are occasional visitors. The marine birds include three penguins as well as albatrosses, petrels, terns, skuas, gulls, a cormorant and a sheathbill. There is also a pintail duck, *Anas eatoni*, peculiar to these islands and Archipel de Kerguelen.

The group was discovered in 1772 by Marion-Dufresne and named by Captain Cook for Julien Crozet who took command of Marion-Dufresne's expedition after the assassination of its leader by New Zealand savages. It is now under French sovereignty. Serious scientific investigation of the islands dates from the establishment, by the French, of a meteorological station on Ile de la Possession in 1961–2.

Archipel de Kerguelen (latitudes 48° 27′–49° 58′S., longitudes 68° 25′–70° 35′E.)

Lying 700 miles east of Iles Crozet is Archipel de Kerguelen, also belonging to France. This extensive archipelago numbers over 300 small islands. The triangular-shaped main island has an area of 2600 square miles. It is characterized by its numerous peninsulas, large bays dotted with islets,

THE AUTHOR, IN HIS SEAL SKIN DRESS.

NARRATIVE

OF A

VOYAGE TO THE SOUTH SEAS,

AND THE

SHIPWRECK

OF THE

PRINCESS OF WALES CUTTER,

WITH AN ACCOUNT OF

TWO YEARS RESIDENCE

ON

AN UNINHABITED ISLAND,

BY

CHARLES MEDYETT GOODRIDGE,

OF PAIGNTON, DEVON,

ONE OF THE SURVIVORS.

FIFTH EDITION, ENTERED AT STATIONERS' HALL.

EXETER:

PRINTED AND PUBLISHED BY W. C. FEATHERSTONE,

AND SOLD BY THE AUTHOR.

1843.

An Antarctic Robinson Crusoe. Charles Goodridge was shipwrecked on Iles Crozet during a sealing voyage in 1819 but lived to publish an account of his adventures

and a coastline deeply indented with fjords. The tableland of the interior, which is volcanic, is dissected by numerous valleys and its jagged ridges and peaks present a striking spectacle from the sea. The highest peak, Mont Ross, attains 6430 ft. There is a permanent ice-cap from which numerous glaciers flow, some reaching the sea; Glacier Cook is the largest. Meltwater streams from these glaciers and an abundant rainfall feed numerous small rivers and lakes. The climate is typically oceanic with a small temperature range averaging 4.5 °C. in summer and 2.6 °C. in winter. Rain or snow is said to fall on between 250 and 300 days of the year and the sky is always partly overcast. The westerly winds are typically persistent, attaining gusts of up to 100 m.p.h. Plant cover follows the usual pattern of the islands in these latitudes, with tussock-grasses predominating at coastal level. Inland, in sheltered places, there are various communities of plant life such as cushion-plants interspersed with Kerguelen Cabbage, valued by the sealers of old as a preventative of scurvy. There are only a

Scientific station Port-aux-Français, Archipel de Kerguelen

few species of flowering plants, but numerous mosses and lichens. Off the shores of these islands drift large patches of kelp or giant seaweed. The wildlife is typical of the sub-Antarctic with Royal, King and Rockhopper Penguins, Elephant Seals, skuas, gulls and terns. The native birds include a sheathbill, a cormorant and a pintail duck (*Anas eatoni*). There are no true land birds. The land animal life is, surprisingly, richer than any of the other southern sub-Antarctic islands and includes spiders, fleas, beetles and even an earthworm.

The islands were discovered in 1772 by a Frenchman, Yves Joseph de Kerguelen-Trémerac. They were visited by Captain Cook in 1773 and by numerous whaling and sealing vessels in the eighteenth and nineteenth centuries. The French now maintain a permanent scientific station at Port-aux-Français.

Heard Island (latitude 53° 06′S., longitude 73° 31′E.)

This roughly circular island, approximately 225 square miles in area, is virtually a mountain rising from the ocean. Its summit is known as Big Ben

Bouvetøya, Kapp Circoncision (*see p. 158*)

from which rises the cone of Mawson Peak, 9005 ft. high. In climate and appearance it is very different from that of Archipel de Kerguelen; permanent ice covers most of the island terminating in ice-cliffs between 50 and 100 ft. high. The average temperature throughout the year is only half a degree centigrade above freezing. Cushion-plants and tussocks are found on a few ice-free slopes but in no great abundance. Mosses and lichens tend to be more widespread. Elephant Seals breed here and the Leopard Seal is a winter visitor. Birds include three penguin species, Macaroni, Rockhopper and Gentoo, as well as petrels, gulls, cormorants and albatrosses. The island was first sighted by a British sealer Peter Kemp in 1833, and again by an Australian, Captain Heard, in 1855. The Australians, who own the island, maintained a weather-station here between 1948 and 1955. In 1965 a private mountaineering expedition from Australia succeeded in climbing Mawson Peak. Near by are the Macdonald Islands, an unexplored rocky outcrop.

4. ISLANDS SOUTH OF NEW ZEALAND AND THE PACIFIC SECTOR OF THE SOUTHERN OCEAN

Lying to the south and south-east of New Zealand are a number of islands whose climates, flora and fauna are closely related to the other sub-Antarctic islands; they include the Snares, Bounty and Antipodes Islands, the Auckland Islands, Campbell Island and Macquarie Island. Except for Macquarie Island, which is under Australian sovereignty, all are administered by the New Zealand Government. Lying on the old shipping route from Australia to Europe via Cape Horn, their rocky fog-bound shores were the scene of many a wreck and there was a time when the New Zealand Government used to send out an annual relief ship to tour them expressly to pick up shipwrecked mariners.

Snares Islands (latitude 48°S., longitude 166° 35′E.)

The southernmost of the New Zealand archipelago this tiny group was first seen by Captain George Vancouver in 1791. They were rapidly

Big Ben, Heard Island's lofty summit. The scientific station, once operated by Australia, is now closed

despoiled of their rich stocks of Fur Seals and by 1810 none were left. In the thick peat which covers parts of the islands nest burrowing petrels and myriads of Cape Pigeons; penguins and other marine birds abound. Elephant Seals are also numerous.

Bounty Islands (latitude 47° 41′S., longitude 79° 03′E.)

Discovered by Lieutenant William Bligh of H.M.S. *Bounty* in 1788 this cluster of small granite islands, 490 miles east of New Zealand, is devoid of vegetation but is a refuge for vast numbers of marine birds.

Antipodes Islands (latitude 49° 41′S., longitude 178° 43′E.)

The group, situated near to the antipodes of London, consists of one large island, named Bollons Island and a number of smaller islands. Bollons Island is mountainous and volcanic in origin. It is largely peat-covered and grows tussock-grass, scrub and ferns. Breeding in the tussock is a small green parakeet which walks and climbs but seldom flies. The islands were discovered by Captain Waterhouse of H.M.S. *Reliance* in 1800.

Auckland Islands (latitude 50° 50′S., longitude 166°E.)

In contrast to the barren rocks of the smaller sub-Antarctic islands of New Zealand, the Auckland Islands display a relatively lush landscape and a greater abundance of animal life. About 200 miles south-west of New Zealand the group consists of several islands of which the largest is Auckland Island, some 345 square miles in area. Other islands include Enderby and Ewing islands to the north, Adams Island to the south and Disappointment Island to the west. The hills of Auckland Island itself, rising to over 2000 ft. in height, are covered with luxuriant vegetation on their lower slopes, the flowers of the Southern Rata (or ironwood) providing great splashes of scarlet. The east coast of the island has several fine harbours, like Port Ross, known more romantically to the sealers as 'Sarah's Bosom'. Some of the smaller islands are of especial interest to biologists since they still retain their original native plants and animals. Many of the plants are so rare that they lack common names. Flocks of parakeets and tiny Auckland Island snipe give the scene an almost sub-tropical air. Considering their relatively high latitude these islands are remarkably temperate and lack any permanent snow and ice. There is, too, a great abundance of marine wildlife. The Fur Seal, once nearly extinct, is gradually recovering, and there are Elephant Seals as well as Giant Petrels, albatrosses, skuas and Rockhopper Penguins which range as far south as Antarctica.

Discovered in 1806 by Captain Abraham Bristow of the vessel *Ocean*, belonging to the London firm of Samuel Enderby and Sons, the Auckland Islands have been the scene of several scientific expeditions. Among the first were those led by Sir James Clark Ross, Dumont d'Urville and Charles Wilkes between 1839 and 1842. In 1849 an attempt to establish a permanent whaling-station on Auckland Island by Charles Enderby was made; but the climate was poor and the soil too little suited to agriculture for it to succeed. The enterprise failed and the colonists were dispersed. Apart from a brief period during the Second World War there have been no permanent stations on the islands.

Campbell Island (latitude 52° 33′S., longitude 168° 59′E.)

This single island lies to the south-east of the Auckland Islands and is about 40 square miles in area. Its hills are not so high as those of the Auckland Islands and are either bare or covered with scrub or tussock. Campbell Island is exposed to the full force of the prevailing westerly winds and is subject to frequent rain and cloud. Temperatures are low though frost and snow are infrequent. Much of the native flora has been destroyed by wild sheep, survivors of a farming experiment abandoned in 1929. Campbell Island has a unique animal and bird population including Elephant Seals, Fur Seals and penguins. The Rockhopper colony here is one of the largest in the world. The island was discovered in 1810 by Captain Frederick Hasselburgh of the *Perseverance*, owned by the firm of Campbell of Sydney, Australia. It has been the site of a permanent meteorological station since the Second World War.

Macquarie Island (latitude 54° 37′S., longitude 158° 54′E.)

This, our last sizeable sub-Antarctic island, lies roughly half-way between Australia and the Antarctic continent. The island is 46 square miles in area, a mountain range rising abruptly from the sea, surrounded by rock ledges on which many a ship has foundered. Macquarie Island must once have been covered with a thick ice-cap; today the island is virtually free of snow during the summer months. Rain or fog occur on over 300 days of the year and the unceasing westerly winds reach speeds of over 100 m.p.h. There are no trees or shrubs but plenty of the kind of vegetation found on other islands in these latitudes – tussock-grass, mosses and cushion-plants, and the remarkable Macquarie Island Cabbage with its rhubarb-like stalks, a foot long, and large waxy flowers. Round the shores float dense carpets of giant seaweed. Macquarie Island has long been a breeding-place for Elephant Seals and King, Royal and Rockhopper Penguins. During the nineteenth century the penguins were slaughtered ruthlessly by the sealers for their oil and narrowly escaped extinction. Only stringent conservation measures by the Australian Government (the island is now a wildlife sanctuary) have saved these birds, along with the Elephant Seals, from extermination. Even so a number of native birds, including a parakeet, have become extinct or are threatened with extinction through the depredations of man and his introduced animals.

Macquarie Island was discovered in 1810 by Captain Frederick Hasselburgh, of the brig *Perseverance*, looking for fresh sealing grounds. It has

The coast of Macquarie Island

Helicopter landing on Sabrina Island, one of the inaccessible Balleny Islands

A sub-Antarctic wildlife sanctuary. Elephant Seals and Royal Penguins at Nuggets Bay, Macquarie Island

been visited by a number of scientific parties, one of the most thorough studies being that of Sir Douglas Mawson between 1911 and 1913. It was he who established the first weather-station on the island. Since 1948 Australian National Antarctic Research Expeditions (ANARE) have maintained 6 permanent scientific stations on the island carrying out research in numerous disciplines.

5. OFF-SHORE ISLANDS OF ANTARCTICA IN THE PACIFIC SECTOR OF THE SOUTHERN OCEAN

Our review of the sub-Antarctic islands is complete; westwards lie the vast and empty wastes of the Pacific sector of the Southern Ocean stretching away until South America is reached. Southward from Macquarie Island across the Antarctic Convergence in colder waters, often covered by pack ice, lie a few islands within a hundred miles or so of the coast of Antarctica which are of historical and scientific interest. They include the Balleny Islands, Scott Island and Peter I Øy.

Balleny Islands (latitude 69° 12′S., longitude 158° 50′E.)

This group of six volcanic islands, about 160 miles off the coast of Oates Land, in Australian Antarctic Territory, are truly Antarctic being thickly covered with ice. They are mountainous, ranging in height from 600 ft. to over 4000 ft. high. Their name commemorates the British sealer John Balleny who discovered them in 1839. Because of their inaccessibility little is known about them. A New Zealand expedition made helicopter landings here in 1963–4 and reported the islands as being quite unsuited for a permanent research station.

Scott Island (latitude 67° 24′S., longitude 179° 55′W.)

This ice-covered island surrounded by steep cliffs lies 310 miles north-east of Cape Adare, Victoria Land. It was discovered in December 1902 by Lieutenant Colbeck of the *Morning* sent to relieve Captain R. F. Scott's *Discovery* Expedition.

Peter I Øy (latitude 68° 50′S., longitude 90° 30′W.)

This remote island lies well within the Antarctic Circle, 210 miles from the Antarctic mainland in the Bellingshausen Sea. It is about 15½ miles long and 6 miles wide and is entirely covered with ice, the bare rock showing only where it is too steep for snow to lie. The highest point is Lars Christensentopp (4003 ft.). The island was discovered in 1821 by a Russian naval expedition led by Captain Thaddeus von Bellingshausen who named it after Peter the Great of Russia. The present Norwegian version of the name commemorates the first landing on the island made from the Norwegian research vessel *Norvegia* in 1929; the island is today under Norwegian sovereignty.

II
POLAR TRANSPORT

Logistics, according to the *Shorter Oxford English Dictionary*, is the art of moving or quartering troops. It is a useful portmanteau word, by analogy, with which to describe the art (some would call it a science) of transporting an expedition and of feeding, clothing and sheltering it with the maximum gain to knowledge and the minimum inconvenience and discomfort to its members. Scientists are increasingly disinclined to spend valuable time in the field doing non-skilled work; on the other hand not all Antarctic expeditions can afford the expensive support which the larger and wealthier nations give to their scientists. In this chapter, and the one that follows, we shall be reviewing some of the logistical problems that arise and the ways in which they are being overcome.

Ships

The bulk cargo needed to keep a scientific expedition in the Antarctic is a formidable one. Provision must be made for a year's supply of food for a hundred or more men (plus a year's reserve), fuel for warmth and cooking, a tractor or two, possibly aircraft, and of course fuel to keep these going in the field. Add to this the sectional parts of a new hut, numerous items of scientific equipment, much of it bulky and requiring special care in packing, plus innumerable items of personal gear and clothing, and it soon becomes evident that there is only one economic way to transport it all south, and that is by ship. As we shall see later one heavy aircraft can transport men and bulk cargoes efficiently and much more quickly but there are various technical reasons why they cannot replace sea transport entirely.

Ships, like humans, must be well adapted to survive in Antarctic conditions. In the chapter on sea ice we gained some idea of the problems facing shipping in the Southern Ocean. We have seen how, for some three-quarters of the year, the sea surrounding Antarctica is frozen solid. During this period no ship can hope to penetrate it. Then for a few brief summer months, from the beginning of December to early March, the edge of the pack ice retreats southwards and in places breaks up to such an extent that ships can push their way through with comparative ease. But even at the height of summer not all parts of the Southern Ocean are navigable; in the Ross Sea, generally considered the most reliable sea-way to the Antarctic, damage to shipping can occur even at this time of the year.

The most common forms of damage are breaking of the propeller blades and rudder, buckling of steel plating and crushing of the hull. From this point of view modern steel vessels are far more vulnerable than the early wooden ships such as Captain Scott's *Discovery*. Powered by steam and sail such vessels were powerfully reinforced to withstand the crushing forces of the pack and were indeed remarkably resilient. But they were very much at the mercy of the ice, being too underpowered to break their way out in an emergency. The classic instance of total loss was the crushing of Shackleton's *Endurance* by the pack ice of the Weddell Sea in 1915. The advent of large-scale scientific exploration in the Antarctic since the Second World War, requiring regular annual relief of stations and the transport of large bulk cargoes, has brought about the evolution of two entirely different types of vessel – the icebreaker and the ice-strengthened ship.

Icebreakers are vessels whose primary task is to break a way through the ice and manoeuvre freely in heavy pack which may be anything up to 20 ft. thick. They are characterized by powerful engines, a sloping bow to enable the vessel to ride up on and smash through the ice, heavily reinforced hulls, and heeling tanks to enable the ship to be rolled from side to side to free itself. Because of the volume of space taken up by the engines there is very little room for cargoes thus an icebreaker's chief function is to break channels in the ice through which conventional cargo ships can pass. In conjunction with these various kinds of cargo ships, oil tankers and amphibious vessels may be used. Of the 12 nations currently active in Antarctica, only Argentina, Japan and the U.S.A. are using icebreakers; other nations would like to do so but the cost is considerable and they therefore prefer multipurpose ice-strengthed ships. Ice-strengthened ships

Japanese Antarctic research vessel *Fuji*

are combined cargo-passenger ships with reinforced hulls and increased engine power. They can navigate moderate pack ice but, unlike ice-breakers, cannot navigate the heaviest pack or break through thick fast ice. The British rely entirely on this type of sea transport, employing two of their own vessels, R.R.S. *John Biscoe* and R.R.S. *Shackleton* for the relief and supply of bases on the Antarctic Peninsula, and chartering a third vessel for the relief of their Weddell Sea base, Halley Bay.

A development of the ice-strengthened cargo vessel is the ship designed specifically for scientific research. The American *Eltanin* is one of the latest examples of this type of polar vessel. *Eltanin*, a modified cargo vessel, is strengthened for ice and is equipped with laboratory space for 35 scientists – she even has anti-roll tanks to prevent them getting seasick.

The hazards of the pack ice surmounted, there is the problem of landing stores on the continent itself. Antarctica has few natural harbours or anchorages and virtually no artificial piers or jetties. Landings have to be made on floating ice or on rock. In some cases ships can tie up alongside an ice-shelf and unload on to sledges which then transport the cargo to base. A novel form of unloading is the cableway, which the French have built at their Dumont d'Urville base in Terre Adélie to lift cargo direct from the ship to the station itself.

Aircraft

To Captain R. F. Scott must go the honour of being the first aeronaut to make an ascent in the Antarctic; on 4 February 1902 he went up some 800 ft. in a captive balloon and saw stretching away in the distance the undulations of the Ross Ice-Shelf. Air transport has revolutionized scientific exploration in Antarctica. With their speed and flexibility aircraft have largely overcome the natural barriers to travel, have made the interior of the continent accessible, and have enabled scientists to devote more time to their science and less to the hard foot-slogging experienced by the old-time explorers. Aircraft are not without their limitations in this environment – there are the vagaries of the weather, the absence of suitable landing-fields, and of course there is the considerable expense involved in supporting them with maintenance crews, spare parts and fuel.

The type of aircraft used in the Antarctic depends on the particular job they have to perform. For long-distance transport of personnel and stores, heavy machines will be needed; for local short-distance runs and air reconnaissance, light aircraft and helicopters are used. The U.S.A., as always in the vanguard of technical development and innovation, has for many years used large aircraft to ferry personnel and stores from their New Zealand base direct to McMurdo Station in Antarctica's Ross Dependency. The 2500 mile trip is covered effortlessly in 8 hours; by ship it would take a week at least, depending on ice conditions. Most versatile of this type of aircraft is the American LC-130 Hercules. Equipped with

Preparing for the first balloon ascent in Antarctica, Scott's *Discovery* expedition, February 1902

skis it has an operating radius of 1500 miles fully laden. The Hercules has the great advantage of being able to land on unprepared snow surfaces inland – a valuable asset in a country where there are few artificial landing-strips. Before the advent of the Hercules inland stations could only be established by dropping supplies by parachute, or by what is called 'free-dropping' – that is jettisoning fuel and supplies in specially prepared containers. Much valuable material was irreparably damaged or lost in this way. The Hercules is equipped with auxiliary fuel-tanks in the fuselage which enable it to deliver fuel to remote locations. The introduction of this aircraft has made it possible to put parties in the field complete with equipment including tractors, helicopters and the fuel to keep them going. Plateau Station, 11,890 ft. up on the remote plateau of Greater Antarctica where the temperature drops to −80 °C., was established in December 1965 by the U.S. Naval Support Force in this way. In the space of three weeks living accommodation for men, and 8 laboratories, consisting of 4 prefabricated huts each weighing 23,000 lbs., 2 10-ton traxcavators to pull them into position, fuel-storage bladders and 50,000 gallons of diesel fuel, had all been flown in by Hercules via the South Pole station. To have accomplished such a feat by tractor train would have been out of the question. Not only the Americans but the Russians, too, with their Ilyushin aircraft have demonstrated the value of this kind of transport.

Unloading fuel drums from a Hercules aircraft at Byrd Station

For local supply and reconnaissance light aircraft such as the de Havilland Beaver and Otter find favour with the smaller scale expeditions. They are rugged and highly versatile but being only single-engined are somewhat 'at risk' over open water. Helicopters are also invaluable in the Antarctic; they can be based on board ice-breakers where they are used

A U.S. Navy helicopter transporting the heavy frames of a sectional hut

to locate leads in the ice invisible from the ship's crow's-nest. Or they may be land-based as general utility aircraft for hauling light cargo, for reconnaissance work or for transporting scientists to otherwise inaccessible working-points such as a glacier or mountain-peak. The helicopter's chief disadvantage is its vulnerability to high winds and its relatively short range (165–300 miles).

The provision of permanent and reliable airfields in Antarctica is attended by all manner of problems. At present the only runways regularly maintained for both wheeled and ski-equipped aircraft are at Williams Field located on the Ross Ice-Shelf near the United States McMurdo Station. Here the runway is built on floating ice which from time to time floats out to sea; hence both runway and air-strip buildings have to be relocated at intervals.

The chief beneficiary of air transport is the scientist; if he hails from an American university he can now fly to the Antarctic in October, do three months research in the field and return to his laboratory by the New Year. Efficient air transport enables the short Antarctic season to be used to maximum advantage.

Helicopter after a crash in whiteout conditions

Williams Field, Antarctica's only semi-permanent airfield, on the floating ice of McMurdo Sound. Ross Island and McMurdo Station in the background

Land transport

Land travel in Antarctica presents problems that are to be found nowhere else in the world. There are of course no roads and surface conditions may be rough, varying with the temperature, the wind and the time of year. To meet these demands a whole family of land vehicles has evolved over the past 60 years. Still in use in the Antarctic is the Nansen sledge pulled by a team of huskies, which for many of us symbolizes the typical form of Antarctic travel. It was Captain Scott who first made extensive use of this form of powered transport during his *Discovery* Expedition to the Antarctic in 1901–4. On his first sledge journey, southwards across the Ross Ice-Shelf, he lost the lot through disease and exhaustion and had to resort to manhauling. Scott again used dogs on his final attempt on the South Pole in 1911–12 but relied on manhauling for the last arduous trek across the polar plateau. Scott also used ponies, but though reared in Siberia they never became wholly acclimatized to the Antarctic and were not a success. Amundsen, Scott's successful rival in the race to the Pole, used dogs with ruthless efficiency, killing off the weaker members of the team and feeding them to the rest. Manhauling of the arduous kind practised by Scott is a thing of the past; but dog-sledging is still practised especially by members of the British Antarctic Survey and the New Zealand and Australian expeditions. For travelling over the kind of weak sea ice often experienced

Dog team and sledge

Husky dog 'Apolotok'

among the islands of the western Antarctic Peninsula, the dog sledge is still by far the safest form of transport. A good dog in the lead will also give his driver due warning of dangerous crevasses. Dog sledges have many other advantages over mechanical transport; there are virtually no mechanical parts to give trouble and a sledge, if broken, is easily mended. The 1 lb. to 1½ lbs. of dried pemmican fed daily to a dog when working can be very compactly packed and stored; and these rations can be supplemented by locally available seal meat when necessary. Fuel for mechanized vehicles, by contrast, is bulky and difficult to supply. A dog-team usually consists of nine dogs hauling a load of up to half a ton. The practice of allocating one driver to each team ensures an understanding between man and dog essential to efficient sledging. In recent years the Australians and New Zealanders have successfully ferried complete dog-teams by air to carry out survey work in remote mountainous areas.

Valuable as dogs are for transporting small parties they cannot compete with mechanized transport where large-scale movement of men and materials is needed. Mechanical transport of the size and scale used on the Commonwealth Trans-Antarctic Expedition of 1955–8 and on United States and Soviet expeditions of recent years provides splendid examples of adaptation to environment. Their history is virtually as old as the internal combustion engine itself. The first mechanical vehicle used in the Antarctic was a 4 cylinder 15 h.p. Arrol-Johnston 200 motor car with an air-cooled engine, using standard tyres, taken down by Sir Ernest Shackleton in 1907–9. This forerunner of the Snocat succeeded in performing a number of short hauls, pulling a sledge, the record being 30 miles in a temperature of −34 °C. The Arrol-Johnston was certainly a great deal more successful than the three Wolseley motor sledges tried out by Scott in 1910–13. These were the first vehicles with tracks to be used in the Antarctic; they were perpetually breaking down and were never of any significant use.

No great advances in the design of vehicles for use in the Antarctic came until after the Second World War. The first effective over-snow vehicle was the Weasel, a tracked vehicle originally designed for use by the United States Army. It has been used successfully by the French on the Greenland ice-sheet and in the Antarctic by the Norwegian-British-Swedish Expedition of 1949–52 and others. The Weasel's relative lack of power and constant need for maintenance make it unsuitable for the needs of present-day large-scale expeditions and it has been superseded by such vehicles as the Canadian Muskeg Tractor, a fast and easily maintained machine. The heavier and more powerful Snocat achieved fame through its successful use by the Commonwealth Trans-Antarctic Expedition and subsequently by the United States 'Operation Deep Freeze' expeditions. The largest version is capable of carrying up to 6000 lbs. of cargo, can tow an additional 15,000 lbs. and has a range of 200 miles. For long-distance

An artist's impression of the first motor tractor used in the Antarctic on Shackleton's *Nimrod* expedition, 1907–9. The front wheels, shod with wood, were provided with sledge-runners. The rear wheels had steel projections fitted, into which holes were drilled to receive spikes designed to increase adhesive power. The engine was capable of giving speeds of up to 16 m.p.h. and the two petrol tanks held fuel for 300 miles

Snocat used by the Commonwealth Trans-antarctic Expedition, 1955–8. The vehicle has a unique traction system enabling it to travel over soft snow or hard ice

Japanese 3-ton diesel snow-cars in the Queen Fabiola Mountains of Dronning Maud Land

Tractor train of Australian National Antarctic Research Expeditions

Soviet artillery tractor converted for use in the Antarctic

A line-up of vehicles at the New Zealand Scott Base

Eliason motor toboggan, driven by remote control

towing the Americans and Australians use Caterpillar D8 tracked vehicles with a maximum horsepower of 385. But even these monsters are dwarfed by the Soviet 12 cylinder 500 h.p. Kharkovchanka vehicle which hauls a 50 ton sledge. The Kharkovchanka is virtually a scientific station on tracks. It can provide working accommodation for 8 men and has a work-room, radio cabin, kitchen heated by an anthracite stove and toilet. In the driver's cabin are the latest navigational aids.

An ideal vehicle for small scientific parties is the small single-tracked motor toboggan. This has proved itself to be a reliable vehicle and can haul a sledge carrying up to 1 ton.

Transport problems

The very low temperatures experienced on the high Antarctic plateau pose a number of problems to the designer. At certain low temperatures metals become brittle, and chassis, springs and tracks are liable to break easily. Successful starting from cold requires pre-heating of the engine and the use of special thin oil. Even batteries may need heating. Constant maintenance is essential to prolong the life and reliability of the equipment.

The difficulties of low temperatures resolved, those posed by the surface must be faced. The snow surface can vary enormously in density according to the state of the temperature and wind. It can be soft and soggy, or it can be rock hard; or it can be dry and granular like fine sand and offer maximum resistance to a vehicle's tracks. This means that considerable thought must be given to the width and construction of the tracks and the distribution of a vehicle's weight. Sledge-runners need to be covered with plastic to lower surface friction.

Sastrugi can bring the mightiest tractors to a halt. They range from a few inches to 5 ft. in height. Iron hard, those over 1 ft. high can damage an aircraft or tear the track off a Snocat. One of the greatest hazards to man and his vehicles is the crevasse. These great chasms in the ice vary from a few inches to over 100 ft. in width and can be big enough to swallow a double-decker bus. The crevasse itself is usually camouflaged by a snow bridge which may be up to 30 ft. thick and it is the easiest thing in the world for an inexperienced or careless driver to plunge into one unawares, all too often with fatal results. The greatest care must be taken when driving through a crevassed area; vehicles can be roped together in twos or threes while a man on skis probes the area ahead with a long stick and marks a safe route with flags. This can be a tediously slow business; progress through a badly crevassed region can be reduced to as little as half a mile a day. United States' scientists have devised an electrical crevasse detector which is pushed in front of a vehicle and can be useful in certain situations. The Americans have also devised a technique of blowing the top off a crevasse with dynamite and then using bulldozers to fill in the chasm with thousands of cubic yards of snow.

Man's ingenuity in the face of nature's challenge seems boundless. His answer to the crevasse and other surface problems may well be the Hovercraft (or Air Cushion Vehicle) which can skim over these obstacles a few feet above the ground and thus gain access to areas which otherwise would have to be approached by dog sledge or on foot.

We have seen something of the enormous progress that has been made in the technique of Antarctic travel. In the beginning not only land travel but the very sea voyage was an enterprise whose outcome was unpredictable. The men who manned the early expeditions had to condition their bodies and their minds to accept the most appalling hardships. It seems incredible that they had any energy or taste for scientific work; yet alongside the bodies of Scott and his companions when they were found by the search-party, were 35 lbs. of valuable geological specimens. Today the picture is vastly different. Speedy and efficient, mechanized transport has opened up all but the most inaccessible parts of the Antarctic continent to the scientist.

12
DESIGNS FOR LIVING

The number of regularly manned scientific stations in the Antarctic now exceeds 40; many, like the British Stonington Island Base in the Antarctic Peninsula, are small – a few scattered wooden huts housing a dozen men and their equipment. Others might accommodate twice this number; only one can claim the status of a township, the United States McMurdo Station, scientific base and transit centre which has wintering accommodation for over 200 men and can cope with over 1000 during the summer relief season. Antarctic stations also vary widely in the materials from which they are constructed and in the range of facilities which they can offer. But this much they must provide in common; reliable protection from the elements, adequate light, heating and ventilation, a water supply and a means of disposing of waste. Basically these requirements are no different to those which we expect an architect to provide in less rigorous latitudes; but as with transport, so with housing, we have to adapt ourselves to a very different environment. The building of a station may have to take place in the severest weather with the thermometer well below zero, perhaps even in the rarefied altitudes of the high Antarctic plateau. The work may have to be carried out with the minimum of auxiliary equipment. Construction teams will be hampered by cumbersome polar clothing, unable to work without insulated gloves. Prefabrication is therefore essential, not only to aid construction, but to enable component parts to be loaded easily into aircraft for transport to the remotest parts of Antarctica. A modern Antarctic settlement is planned down to the minutest detail and the method of construction carefully rehearsed in advance to ensure that the actual work of assembly is as troublefree as possible.

In some respects basic hut design has not altered fundamentally since the days of Scott and Shackleton whose wooden huts on Ross Island still stand. They are preserved as historic monuments by the New Zealand Government. Advancing technology can now supplement wood with many

Stonington Island, a British station in Marguerite Bay, Antarctic Peninsula

new light-weight and fireproof materials such as aluminium and plastics; more emphasis too is placed on the provision of a homely atmosphere and a greater degree of individual privacy than in the past. In fact 'house' would be a term more acceptable to polar architects than 'hut'.

The design and arrangement of a station is essentially dependent on the site and local climatic conditions. A rock foundation is the best but there is an acute shortage of such snow-free sites in the Antarctic. The alternative is to build on the ice surface. This has numerous disadvantages from an engineering point of view. As we saw in a previous chapter, the ice-sheet which covers the Antarctic continent is constantly on the move; anything built on it will also move, slowly but surely, seawards. A hut built on the ice will also become drifted up by falling snow, or by snow blown by winds, and in time will disappear completely. This has happened at the British station Halley Bay, a medium-sized scientific station built on an ice-shelf of the Weddell Sea in 1956. Snow accumulating at the rate of 3 ft. a year completely buried the main hut to a depth of 30 ft. and the entire station had to be rebuilt. Drifting up is only one of the hazards;

Fireproof building at Soviet station Molodezhnaya, Enderby Land

Construction work on the new British station at Halley Bay

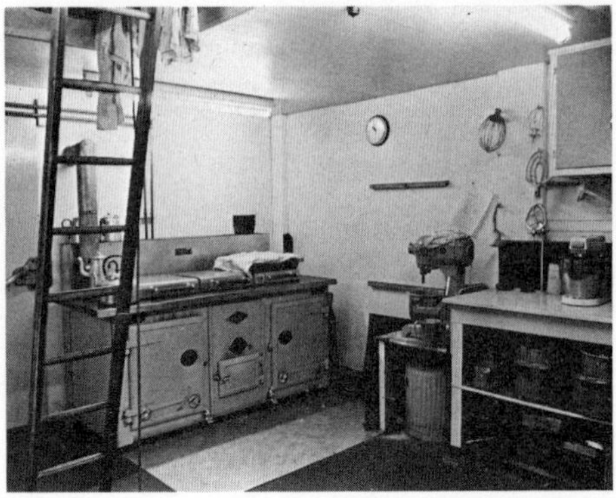

Kitchen, Halley Bay Station

Dining-room, Halley Bay Station

Lounge, Halley Bay Station

New Zealand's Scott Base, Ross Island, McMurdo Sound

Chapel of Our Lady of the Snows, McMurdo Station, Ross Island, McMurdo Sound

General view of McMurdo Station. The nuclear power plant is half-way up Observation Hill in the background

Street scene, McMurdo Station

heat loss through the foundations will cause the underlying ice to melt and a whole building will sink slowly downwards. Pressure of snow on the roof bearing down on the walls will cause the outer edge of the foundations to sink more rapidly than the centre. Thus the walls sink and the floor is distorted.

A complete answer to the problems of building large stations on snow has yet to be found. One solution to date has been tried by United States Army engineers who discovered, after experiments on Greenland's ice-sheet, that prefabricated huts could be erected and maintained for a long period in snow-tunnels. The techniques used are comparatively simple and rely largely on local material – namely snow. First a trench is excavated in the snow using a machine called a Peter Snow Miller. This 15 ton excavator has hydraulically driven rotor blades which can cut a trench in the snow 9 ft. wide, 3 ft. deep and 300 ft. long in half an hour, blowing the snow clear of the trench through a chute. The trench is then roofed over with steel arches which are covered by snow to ground-level. Byrd Station*, situated on the ice-sheet of Byrd Land, provides a good example of this method of construction. Built in 1961–2 to replace the original Byrd Station, then on the verge of collapse from an accumulated weight of snow, it consists of a main trench 600 ft. long, 15 ft. wide at the top and 21 ft. wide at the base, off which branch a number of side trenches. Arranged along these tunnels are standard prefabricated huts insulated from the ground by wooden pads and providing living quarters, laboratories and workshops. There are very many advantages to this type of construction. The huts themselves are completely sheltered from wind and snow and can therefore be built of less robust materials than those used on the surface. Complete protection is also given to power and communication cables which can be secured to the walls of the tunnels. The tunnels enable people to move safely and easily round the station and also provide ample storage space. Utopian as these subterranean cities may seem, their life is inevitably limited; the ice all round is constantly moving and the walls of the trenches will tend to encroach on each other; a temporary solution is to shave them down periodically. Danger from fire, too, is far greater than on the surface. And of course the method requires expensive equipment and is really only suitable for larger stations.

If you cannot afford to dig down then one solution is to mount your huts on skis and move them round when snow-drifting threatens their safety. This method has been used in the construction of the United States' Eights Station. Another solution is to support the huts on telescopic stilts which can be gradually jacked up, a method tried with radar stations in Greenland but not so far tested in the Antarctic.

Engineering marvels such as Byrd Station are scarcely typical; the smaller stations, whether on ice or rock, consist of clusters of huts or

* Sometimes called New Byrd Station to distinguish it from the original installation.

Laboratory buildings at French base Dumont d'Urville

specialized scientific buildings. Huts vary greatly in design and materials according to their functions and architectural preference. Modern British huts may be of steel frames clad with timber, or of prefabricated fibre-glass laminate supported by a steel frame, such as the living hut at Deception Island or the living hut-cum-laboratory at Signy Island. The Americans preferred a more light-weight form of construction for their Plateau Station, which had to be airlifted to the high polar plateau; these huts are basically insulated boxes or vans of plywood covered with aluminium and constructed in units so that they can be joined together with covered passages.

But whatever the materials or method of construction, surface-built huts in the Antarctic must be soundly constructed and well anchored to withstand the elements. Windows, if provided, must be double or even treble glazed to reduce heat loss. An entry porch is essential to keep out wind, prevent heat loss when the door is opened, and provide a place where

British station at Signy Island, South Orkney Islands. The circular tank (*right centre*) is used for storing meltwater

Left: Empty barrels used to support the ceiling of old Byrd Station were eventually crushed by the accumulated weight of snow pressing on the building

Right: A 19-ton Peter Snow Miller used to cut trenches for the new Byrd Station

Below: Smoothing the sides of the trench walls with scrapers

Left: Installing sections of steel arch above a trench

Right: Completed tunnel

Below: A prefabricated building erected in a tunnel stands 3–4 ft. off the tunnel floor to prevent heat transfer and subsequent melting of the snow

snow-covered clothes can be hung out to prevent them thawing and becoming wet. Emergency exits in case of fire, and trapdoors in the roof as a means of escape if you get drifted up, are basic elements in hut design. A rugged exterior, however, can sometimes conceal an interior with every domestic amenity. Thus the van units of Plateau Station provide curtained bunks with spring mattresses for its scientists and have colour combinations carefully selected to create a pleasant, unobtrusive atmosphere.

The public utility services which we take so much for granted at home, such as electricity, water and waste disposal, all present their own special problems when translated to the Antarctic. Heating an Antarctic station presents a number of difficulties. Electric power is clean, convenient and reduces danger from fire; unfortunately electricity is usually at a premium at many of the smaller stations and a variety of other methods of heating is used which include ducted hot air as hot-water systems tend to freeze up. Plateau Station uses an ingenious system of heat exchange, utilizing spare heat generated by the station's diesel generators. An added luxury here is a humidifier included to add water vapour to the virtually moistureless Antarctic air.

Water supply in Antarctica is never a problem providing you have a heat supply to melt the snow. Unfortunately, a great deal of energy is needed to melt enough snow to provide water for all purposes and as energy means oil, and this is expensive, most small stations have to do without the luxury of nightly hot baths. On a small scale waste heat from generators can be transferred to snow-melters into which clean snow must be regularly shovelled. On a larger scale a deep shaft may be drilled into the snow to a depth of 200 ft. or so with the aid of a steam jet; as the snow melts the water is pumped out.

The disposal of rubbish and sewage is a relatively simple matter on a small station – it can be sealed into empty oil drums and left on the sea ice. Complications arise on the larger stations, especially rock-based townships like McMurdo, where more sophisticated systems involving

Control panel of the nuclear power plant at McMurdo Station

A cache of 45-gallon drums of fuel oil – the life blood of a modern Antarctic expedition – used for heat, light and power

heated and insulated sewage-pipes must be provided.

The size of an Antarctic station, if it is to provide a standard of protection and comfort comparable with conditions at home, must bear a direct relationship to the power supply available. The fuel oil normally necessary to provide this energy is bulky and must be transported from a base thousands of miles distant – it is therefore extremely costly. A nuclear reactor, by contrast, can supply a vast amount of energy with little re-fuelling. Such a reactor has been driving electric generators at McMurdo Station since 1962 providing electricity and heating. Now it is operating a desalinization plant to remove salt from the sea and thus provide an abundance of fresh water for the station.

Clothing

Man has contrived for himself in the Antarctic an artificial environment in which he can exist at a level of physical comfort as high – in some ways higher – than that which he would enjoy at home. Temperature and humidity are nicely balanced, internal décor is relaxing and the kitchen service, as we shall see, would satisfy the standards of many guides to good eating at home. But you cannot explore the Antarctic from the comfort of a scientific station and when out in the field man is at once confronted by those two major elements of the polar climate – low temperature and high winds. It is the combination of these two factors which determines the cooling power of the air. Temperature alone will not give an adequate answer to the question: 'How cold is it?' On a fine summer's day at a coastal station in the Antarctic, when the sun is shining and the air is still, there is nothing more pleasant after a good lunch than to lie out on a blanket and sunbathe bare to the waist; the thermometer may register below zero but you do not feel any sensation of being cold. A breath of wind, however, will send one scurrying indoors. It is useful to be able to know just how much body heat will be lost with a given combination of air speed and temperature and this can be done by multiplying a factor for wind, which gets bigger as the wind increases, by a factor for temperature which gets bigger as the temperature drops. The result can be expressed as the 'wind-chill factor'. Prevention of heat loss due to wind chill is one of the major problems of polar clothing design. For wind-chill can nip any unprotected part of the body – fingers, nose, ears and feet are especially vulnerable – and produce frostbite, a damage to tissue akin to a burn in which the protective skin is killed and puffs up like a balloon. Unless circulation of the damaged area is rapidly restored gangrene can set in and amputation may be necessary. Prolonged chilling of the 'core' of the body will lead to a state of lethargy followed by death. Efficient clothing therefore has not only to keep out the wind but must also conserve body heat. One way of keeping warm is to pile on more clothing but too much of this makes movement difficult and stops you working efficiently. Also overclothing causes excessive perspiration; this will then condense to water, freeze within the fabric of one's clothing and thereby destroy its insulating properties. This tendency used to be a major bugbear on Antarctic expeditions in the days before man had begun to study himself in a cold environment. Edward Wilson writes in his diary for 1903: 'During the day all the perspiration freezes in your outer garments under the burberry garments which are impervious and macintosh to keep the wind out. No perspiration gets through this, but you feel dry because it all freezes. But once in your sleeping bag, it all gradually gets warmed through, and every bit of clothing gets damped *very* damp, and worse every night as the days go on. There is no possibility of ever getting anything dried, or of ever getting any clothes off, except socks, which have to be changed night and morning to avoid frozen feet'.

The conflict between the need to insulate the body against chill and at the same time allow it to 'breathe' must result in a compromise. And polar clothing, which has to cater for all man's needs, ranging from racing after a team of exuberant sledge dogs to bending over a theodolite for long periods on a survey traverse, is very much a compromise in design. A

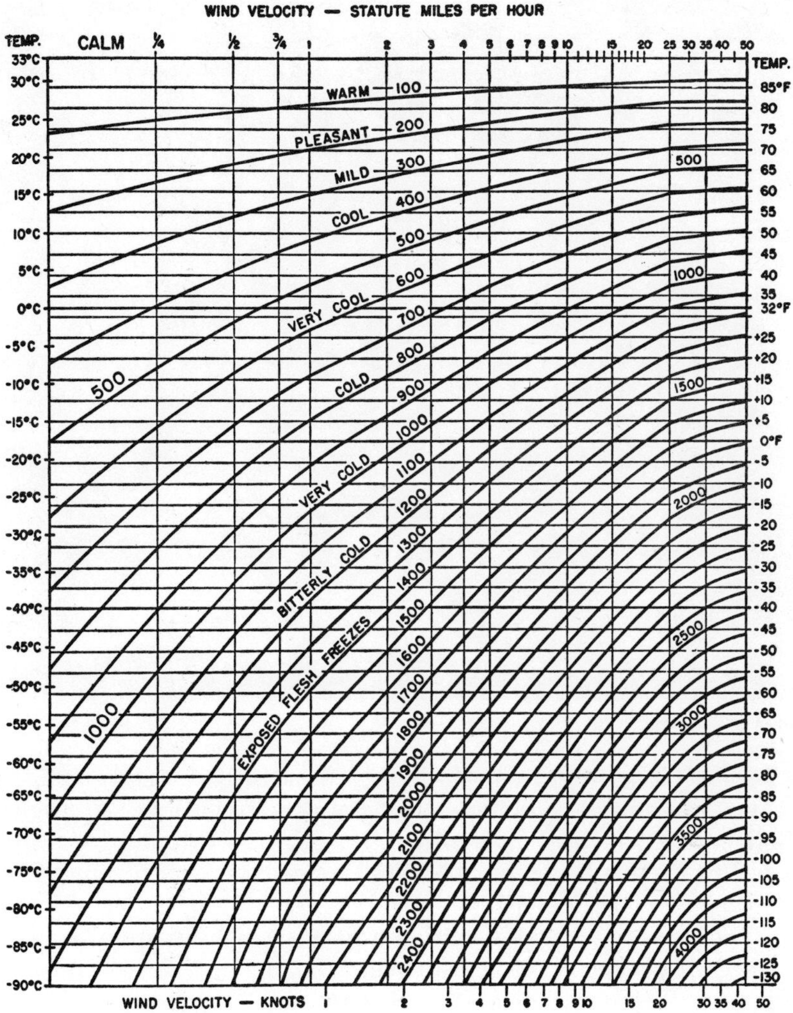

Wind-chill index. The wind-chill factor is an index expressing the relative loss of heat from a heated body and is determined from a formula developed by Dr. Paul Siple. To use the chart, find the estimated or actual wind speed in the top row (miles per hour) or the bottom row (knots). The wind-chill factor is found where the two intersect. With a temperature of –40° (the point at which Centigrade and Fahrenheit temperature scales coincide) the wind-chill factor is 1,400, at which exposed flesh freezes. (Reproduced from Paul Siple's *90° South*, 1959)

well-designed polar suit must conform to the following basic requirements:

1. It must insulate the body from cold.

2. It must be easy to regulate for varying climatic conditions and varying operational requirements.

3. It must be light-weight and comfortable.

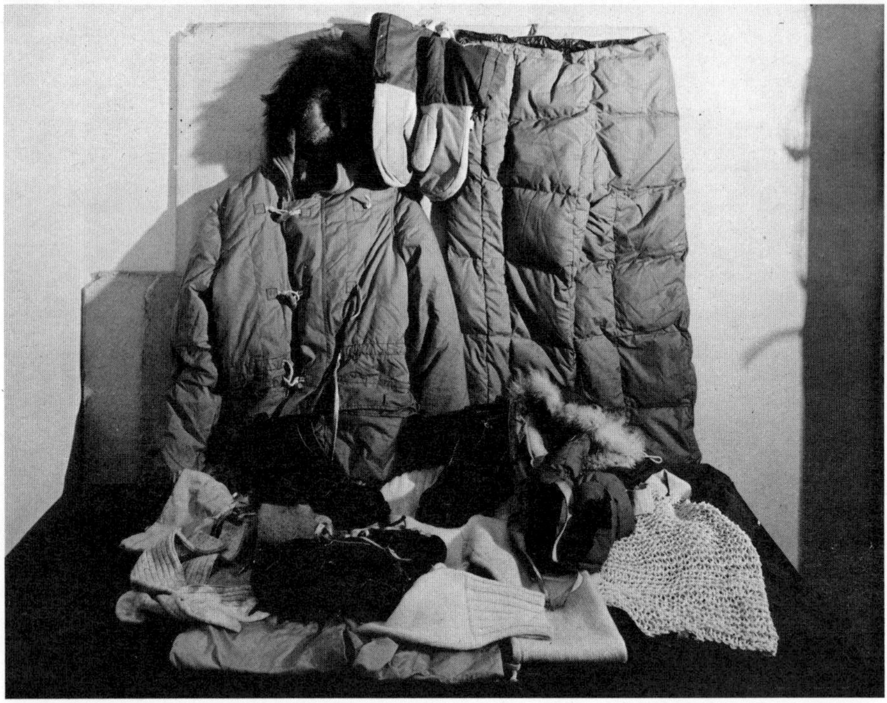

Selection of Antarctic clothing

Now let us look at what the well-dressed scientist is currently wearing in the Antarctic and note how his clothing aims to fulfil these requirements. His suit is based on the layer principle; instead of donning one heavy garment, protective layers are added as conditions worsen. First, next to the skin a two-piece undershirt and pants (or long johns) is worn. This underwear is made of 'waffle-knit' cotton or woollen material creating a number of air-pockets on either side which enable the fabric to trap warm air, the best insulator known. Over this are worn windproof trousers and jacket (or parka) of cotton or nylon. Both trousers and jacket may be supplied with detachable liners of synthetic fabric, which can be used to increase or decrease insulation as required. Both will be baggy, again to ensure the maximum of trapped air which will not only insulate but also absorb moisture due to sweating. Attached to the jacket is a hood trimmed with fur and lined with flexible wire so that it can be shaped to protect the face from wind. Underneath the hood is worn a warm cap. The hands are protected by outsize insulated mittens with leather palms. Most difficult of all to protect are the feet which tend to sweat heavily. Wet socks can easily lead to a breakdown in insulation and this in turn to frostbite. The problem has been solved by the use of the 'thermal boot' which is completely waterproof; the sweat simply accumulates in the boot. The

Typical Antarctic outfit worn by members of United States Antarctic Research Program

feeling of walking about in wet feet takes a bit of getting used to, but since the boot is insulated no heat is lost and therefore no harm can result. All that is required is a daily drying of socks.

This example of a polar suit – and of course there are numerous variations and continual improvements – satisfies all the basic requirements we have listed. Its total weight is around 15 lbs. compared with the 40 lbs. odd of clothing which used to muffle up the old-time explorer. The materials from which it is made are easily washable and non-shrink; even the wool has had the tickle taken out of it!

These basic principles of protection and comfort were discovered through centuries of hard experience by the Arctic Eskimo long before scientists worked it all out from basic theory. Unfortunately, the animal fur, which the Eskimo uses to such effect, is nothing like sufficient in quantity to supply the present Antarctic population. But undoubtedly the type of clothing we have described, using synthetic materials, can keep a scientist alive and comfortable under conditions which would confine a prudent Eskimo to his igloo.

What of the future? The Russians and Americans have done some experimental work with air-conditioned suits electrically heated (one such has two fans to circulate air round the wearer!). But such suits are not yet *de rigueur* in the far south. It seems unlikely that a single all-purpose suit providing complete protection in all conditions of work and climate will be available in the immediate future.

Food

Fifty years ago the success or failure of an Antarctic expedition would have been decided by those twin scourges of polar exploration – scurvy and slow starvation. Scurvy, a deficiency disease caused by lack of vitamin C in the diet, was partly responsible for the death of Scott and his companions on their long march back from the South Pole in 1912. Since that time much research has been undertaken into human energy expenditure and food requirements in the Antarctic. Physiologists are still uncertain as to whether the large appetites induced by Antarctic cold are the result of a change of climate or increased muscular activity, or both. Nor are they agreed on what constitutes an ideal polar diet; undoubtedly more calories are required to compensate for the extra energy expended but just how many calories will vary very much from individual to individual and how active he is. Similarly the composition of polar rations has varied over the years; a satisfactory daily balance for a man working out-of-doors would appear to be 20 per cent protein, 40 per cent fat and 40 per cent carbohydrate providing an energy output equivalent to that needed by a professional footballer. A comparison between modern sledging rations with those of Captain Scott demonstrates the greater variety of food available today, important dietetically and psychologically. That old standby, pemmican, nutritious and compact, but to many simply nauseating, has virtually vanished. Now, processing and packaging techniques allow a modern sledging-party to enjoy many of the luxuries once found only at a base station – freeze-dried meats can be reconstituted in a matter of minutes by soaking in water. As for the food available at the larger base stations, like McMurdo, this would bear comparison with that on the menu of any respectable hotel at home thanks to the fresh meat and deep-frozen fruit and vegetables brought south in specially refrigerated ships.

Communications

Of all the technological advances that have helped man in the gradual process of successful adaptation to living in the Antarctic surely radio communication is the most striking. In the early days of exploration an Antarctic expedition would be completely cut off from contact with home once the ship had left its last civilized port of call. No messages could be received or sent out until a relief ship came or the expedition returned. The news of the disaster which overwhelmed Scott and his South Pole party reached London in February 1913, three months after the bodies were found on the ice-barrier in November 1912. Today radio narrows the distance between a scientist in the Antarctic and his family at home, provides navigational aids for aircraft and a listening service for distress calls; it enables a field-party to maintain contact with base and, if need be, with supporting aircraft. By means of radio small meteorological stations round the Antarctic continent transmit weather data to one or two 'mother stations' who in turn relay it to an analysing centre in Australia where it is available to meteorologists all over the world for forecasting and other purposes.

Unfortunately for those who have to organize this complexity of radio services the Antarctic is bedevilled by the most difficult radio conditions. Owing to the presence of the auroral zone 'blackouts' are frequent and no radio communication is possible. Or interference from static electricity may be caused by drifting snow blowing across aerials during blizzards. The problem of earthing electrical equipment at Antarctic stations causes a great deal of local interference which is difficult to eradicate. No satisfactory solutions have yet been found to any of these problems.

The art (or science) of Antarctic logistics has made great strides in recent years taking advantage of every advance in modern technology. Scientists can live and work in the Antarctic in a greater degree of comfort than ever before; but the degree is a relative one. No one who really knows the Antarctic has yet suggested that scientists might bring their families with them. Few would want to do so. There is a long way to go before the Antarctic becomes ripe for civilized development. Sea transport must be made more reliable and permanent harbours built. More airfields are needed and communications must be improved. Transport has yet to be perfected and such natural hazards as crevasses overcome. More thought needs to be given to clothing design and nuclear energy needs to be made available on a wider scale. Perhaps when these requirements have been met the Antarctic will cease to be merely a scientific laboratory and emerge as man's first internationally organized group of permanent human settlements on Earth.

There is a danger that any simple account of man's technological progress in Antarctica may help to breed a legend that, like some war-time beach now cleared of mines, Antarctica is 'safe' for all comers. Nothing could be further from the truth; all the old dangers are there and no amount of science can save a man from the folly of refusing to learn from the experience of others and not observing the rules. Travel anywhere in Antarctica is potentially dangerous and one should never travel singly. It is possible to get completely lost within a few yards of a base if a blizzard blows up without warning or whiteout conditions descend. In the field, travel is beset with all manner of dangers, the chief being crevasses. Starvation is another as, except near the coast, there is no natural food supply. Sledging on sea ice requires constant alertness as square miles of it can quite suddenly break away from the land and drift out of sight. Most lives are lost because people neglect to follow the few simple rules laid down in the field guides and manuals issued by most countries operating in the Antarctic.

'Three men inside a pyramid tent.' A pencil drawing by E. A. Wilson. The tent is still an essential item of field equipment but camp life is far less spartan than in Captain Scott's day, thanks to scientifically designed clothing and balanced rations

Soviet station Vostok

Soviet station Mirny

Australian station Mawson

13
THE UNVEILING OF THE ANTARCTIC

Rather more than 50 years have passed since Captain Robert Falcon Scott achieved the South Pole to be confronted by a Norwegian flag, silent testimony to the presence there, only a few weeks before, of his rival Roald Amundsen. Today the masts and antennae of Amundsen–Scott Station break the limitless horizon of the polar plateau at 90° south. Powerful aircraft effortlessly scale that *via dolorosa*, the Beardmore Glacier, on routine flights from McMurdo Station to the Pole, symbols of man's increasing mastery over this formidable environment. With the geographical exploration of Antarctica virtually at an end and the major features on the map filled in, future discoveries will largely be in the realms of science. The old-style polar expedition led by figures of heroic dimension, men like Scott, Shackleton, Amundsen and Byrd, has given way to a permanent occupation of the continent by highly skilled scientists and technicians. The story of the early Antarctic expeditions is an absorbing one. We cannot hope to understand the Antarctic of today without a backward glance into the past. We need to make some attempt to follow the current of events which carried men slowly and inevitably towards this most challenging of the Earth's natural regions. All the usual motives are there – commercial advantage, political one-upmanship, private ambition – as well as a somewhat greater than usual appetite for fresh knowledge.

The beginnings

The discovery of Antarctica begins properly with Captain James Cook's second great circumnavigation of the world in high southern latitudes between 1772 and 1775. But its roots go back much further into time, 2000 years and more, to the days when the early Greeks, assiduous star-gazers, named the brightest of the constellations circling the celestial vault *Arctos*, the Bear, and the pole about which it appeared to turn the Arctic pole. The need which the Greek philosophers felt for a balance in the natural order logically demanded a similar pole in the opposite heavenly vault – an anti-Arctic pole in fact. The idea of a celestial Antarctic thus preceded that of a southern land mass. An Antarctic continent would indeed have been inconceivable to the early Greeks who believed in a flat Earth the underside of which was a region of eternal darkness. Not until Aristotle's revolutionary demonstration that the Earth was a sphere did the theoreticians set to work to fill the new Southern Hemisphere with imaginary lands and seas. None of these could they have hoped to explore for an inhibiting theory of climatic zones caused the Greeks to believe that the temperate Mediterranean, centre of their world, must be separated from a temperate region in the Southern Hemisphere by an impenetrable torrid belt of all-consuming heat. To the south of this barrier they believed there must be a frigid zone to counterbalance the frozen regions of the Northern Hemisphere, visited about 320 B.C. by the Greek navigator Pytheas who may well have sailed as far to the north as Iceland.

Early discoveries

The eclipse of Greek scholarship during the early Middle Ages meant an end to both speculation and exploration. To the early Christian Fathers talk of an inhabited antipodes was seen as both idle and heretical. Not until the revival of Classical learning in Europe, which coincided with the dawn of modern geographical exploration, could the discovery of Antarctica become a possibility. Then, one after another, the old barriers of prejudice and superstition fell. The southward probing of the Portuguese Prince Henry the Navigator, between 1406 and 1460, and his successor Bartholomew Diaz, down the east coast of Africa culminating in the discovery by Vasco da Gama in 1497 of a sea route to India round the Cape of Good Hope, put paid to any lingering doubts as to the existence of an impassable torrid zone with seas infested by bizarre sea beasts. And then discovery followed on discovery.

In 1492 Columbus made his eventful voyage to the West Indies to be followed by the Florentine Amerigo Vespucci in 1501, who conceivably voyaged as far south as Patagonia. In 1520 the Portuguese navigator, Ferdinand Magellan, while circumnavigating the globe, passed from the Atlantic to the Pacific along the strait which bears his name. The land to the south of this strait he named Tierra del Fuego, 'Land of Fire', and reckoned it to be a series of islands. The geographers and map-makers at home, however, were wont to interpret these new discoveries in numerous, and often fanciful, ways. A typical world map of the period, that of Oronce Finé (Latinized as Orontius) published in 1531, shows Tierra del Fuego as the northern tip of a great continent centred on the South Pole and filling much of the Southern Ocean. The continent bears the legend 'Terra Australis recenter inventa sed nondum plene cognita' – 'The southern land newly discovered but not yet fully known'. Finé's map is one of the earliest to show the supposed southern continent and first to

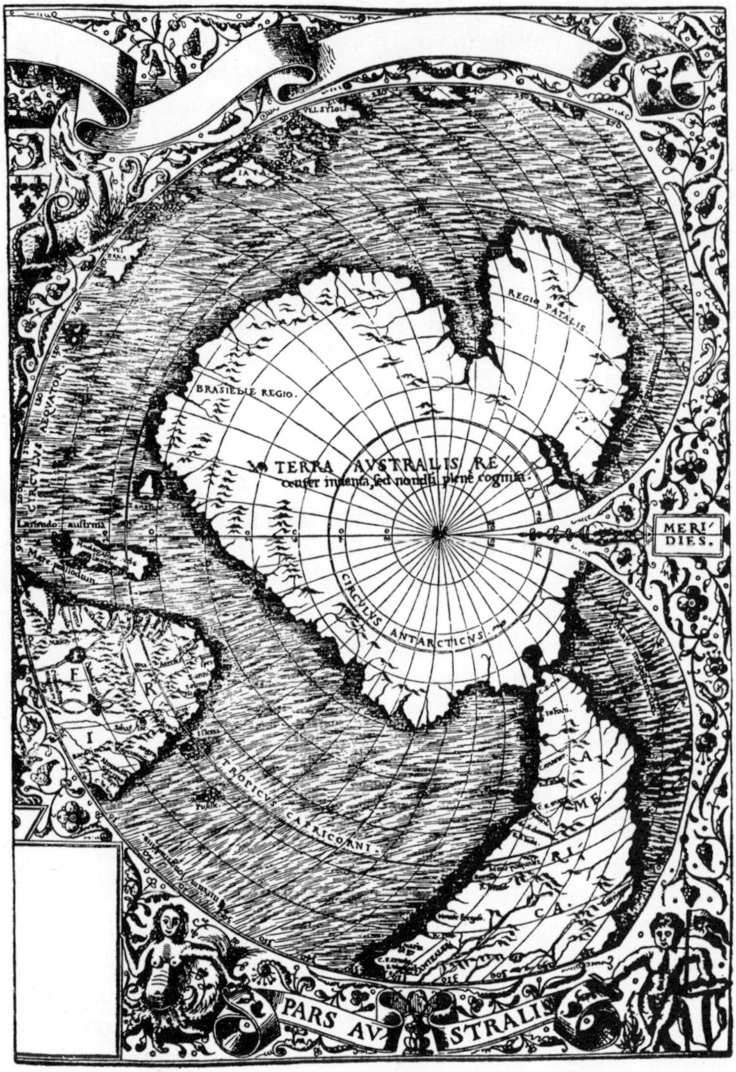

Oronce Finé's map of the Southern Hemisphere, 1531. One of the earliest maps depicting the supposed southern continent surrounding the South Pole, it was the first to use the name Terra Australis

use the name 'Terra Australis'. Many Renaissance cartographers, including the great Mercator, featured the continent on their world maps justifying its presence on the grounds that the land masses of the Northern Hemisphere must be balanced by a corresponding weight of land in the south. Subsequent sightings of land in the Pacific Ocean were seized upon by the map-makers to extend Terra Australis as far north as the Tropic of Capricorn, and even beyond, so that by the latter years of the sixteenth century the southern continent had assumed an area greater even than the two Americas. The search for this mythical land and its hypothetical riches drew in turn a succession of navigators further and further to the south. Their reports all too often showed the armchair explorers to be in error. In 1578 Francis Drake, in the *Golden Hind*, was carried by high winds to the south of Cape Horn into those open seas between South America and the Antarctic Peninsula which we today call Drake Passage. This Drake described as a region where 'there is no maine or iland to be seen to the southwards, but that the Atlanticke Ocean and the South Sea meete in a most large and free scope'. This, the first truly Antarctic discovery to be recorded, in no way caused conventionally minded geographers to alter their steadfast and erroneous course. Nor, indeed, were their theories affected by the Spanish discoveries of the Solomon Islands, the Marquesas and the New Hebrides in the South Pacific between 1557 and 1605, nor by those of the Dutch whose expeditions under Tasman, between 1642 and 1644, showed that Australia was a large island of continental proportions.

James Cook

'The ice islands.' From James Cook's narrative *A voyage towards the South Pole . . .*

Notwithstanding all these major amputations, Terra Australis survived until well into the eighteenth century when it became a potentially valuable objective in the eyes of the great colonial and trading Powers of the day. The British Government promoted several voyages with the aim of developing trade with a land of which one enthusiast wrote: 'the scraps from this table would be sufficient to maintain the power, dominion and sovereignty of Britain by employing all its manufactures and ships'. As late as 1763 world maps were still showing the coastline of Terra Australis as depicted by Finé. Not until Lieutenant James Cook's first voyage round the world in 1769–70 showed that New Zealand was an island, and no continental extension, did it become increasingly evident that a southern continent, if such a continent did indeed exist, must be a very different place in scale and nature from the legendary Terra Australis.

Antarctic exploration begins in earnest

In 1772 Cook, with two Whitby-built colliers, refitted as naval sloops, the *Adventure* and the *Resolution*, was dispatched by the British Admiralty with the express object of settling the problem for once and for all by circumnavigating the globe as far to the south as possible. Cook had some inkling of what to expect, for as long ago as 1739 a Frenchman, Jean-François-Charles Bouvet de Lozier, had discovered and described icebergs and penguins in the South Atlantic at latitude 48° 50′S., and was the first person to report ice-covered Bouvet Island (now Bouvetøya). In the very year of Cook's departure another Frenchman, Yves-Joseph de Kerguelen-Trémarec, had reported the sighting of 'South France', a land rich in economic promise, which, on a closer inspection a year later, was revealed as a cold and barren island group, the present Archipel de Kerguelen.

Cook's experiences in the pack ice are fully described in his *Voyage towards the South Pole*, first published in 1777 and subsequently reprinted in numerous editions and in many languages. Cook and his companions were the first Europeans to cross the Antarctic Circle and to demonstrate that there was no continental land north of latitude 60°S. in the Indian and Atlantic Oceans. Cook's cruise in the South Pacific convinced him that 'the greatest part of this southern continent (supposing there is one) must be within the polar circle', though his later discovery of South Georgia in 1775 led him to the considered opinion that to 'judge of the bulk by the sample it would not be worth the discovery'. It was during the second season of cruising in 1773–4, that Cook reached his furthest south in latitude 71° 10′S., almost within sight of the present Walgreen Coast of Byrd Land. The record remained unbeaten in that longitude until as recently as 1959–60, when two United States icebreakers forced their way into the area. Cook's final discovery, which he named 'Sandwich Land', led him to conclude that 'what we had seen . . . was either a group of islands or else a point of the continent. For I firmly believe that there is a tract of land near the pole which is the source of most of the ice that is spread over this vast Southern Ocean'. His belief in an Antarctic continent was justified, as we now know, though 'Sandwich Land' was in fact one of the Southern Thule group of the South Sandwich Islands. Cook cannot strictly be said to have discovered Antarctica but he was responsible for removing from the map the old imaginary Terra Australis and thereby scaled the southern continent down to a probable ice-covered land. With its clearly defined objectives, its scientists and its well-founded ships this was the first Antarctic expedition in the modern sense. It had many additional advantages over its predecessors, including the newly designed marine chronometer which made the correct determination of longitude possible for the first time. Above all it had a captain whose accurate charts bear comparison with those of the present day and whose solicitude for the health and welfare of his crew was to set high standards for future Antarctic navigators.

Not until 45 years after Cook's voyage did a comparable expedition enter Antarctic waters. Led by Captain Thaddeus von Bellingshausen of the Imperial Russian Navy, it set sail in 1819 from St. Petersburg with the blessing of the Emperor Alexander I and orders to make discoveries 'as close as possible to the South Pole'. Between 1819 and 1821 Bellingshausen's two ships, the *Vostok* and the *Mirny*, completed a remarkable circumnavigation of the Antarctic continent, complementing rather than repeating the discoveries made by Cook. From South Georgia, whose southern coast he thoroughly surveyed, Bellingshausen sailed round the Southern Ocean in a clockwise direction. It was he who showed Cook's 'Sandwich Land' to be no part of the mainland and who, in 1820, viewed from afar the mountains of present-day Kronprinsesse Märtha Kyst in

Thaddeus von Bellingshausen

Dronning Maud Land. By January 1821 the Russian ships had almost sailed full circle when, in the sea that today bears Bellingshausen's name, Peter I island (Peter I Øy) was discovered, to be followed by the sighting of Alexander I coast in the south-west corner of the Antarctic Peninsula, now known as Alexander Island. Despite these landfalls Bellingshausen never claimed to have discovered the Antarctic mainland. In February 1821, having charted and mapped the shores of the South Shetland Islands, he turned the *Vostok* and the *Mirny* northwards for home. Bellingshausen's great contribution to Antarctic discovery was never recognized by his contemporaries and Russia herself took no further interest in the region until as recently as 1956, when the Soviet Government established scientific bases on the continent as part of their contribution to the International Geophysical Year of 1957–8. Today the Soviet Union is a major participant in Antarctic science and the significance of Bellingshausen's discoveries fully recognized.

The sealers

These Russian discoveries in the region of the Antarctic Peninsula coincided in time with those of British and American sealers. Commerce followed closely in the wake of discovery when Captain Cook's reports of Fur Seals on South Georgia caused a major 'fur rush' in the area. By 1820 the sealers were plying their gory trade as far south as the South Shetland Islands, discovered by a British sealer, William Smith, in 1819. In January 1820 Edward Bransfield, a British naval officer, accompanied Smith on a return visit to the South Shetlands and took possession of them for the Crown. On 30 January Bransfield sighted the north-western coast of the Antarctic Peninsula which he named 'Trinity Land'. Bransfield has the distinction of being the first to chart a portion of the Antarctic mainland. His name is commemorated by the strait that separates the South Shetland Islands from the southernmost tip of the peninsula.

To the south of the South Shetlands the islands known as the 'Palmer Archipelago' remind us of the discoveries of Captain Nathaniel Palmer, an American sealer from Stonington, Connecticut, who viewed the distant mountains of the Antarctic mainland from the heights of Deception Island in November 1820. Later, Palmer joined forces with a British sealer named George Powell and together they discovered and charted the South Orkney Islands in December 1821, Powell taking possession of Coronation Island in the name of George IV.

Between 1822 and 1824 the seas east of the peninsula were explored by a Scottish sealer, James Weddell, who found open water as far south as latitude 74° 15′S. in longitude 34° 16′W. Weddell named this new discovery the 'Sea of George IV'. We now know it as the Weddell Sea, a region notorious for the severity and impenetrability of its ice. Weddell's own record in this area has yet to be beaten.

First views of the South Shetland Islands reproduced from an account of William Smith's voyages in 1819 by John Miers, published in the *Edinburgh Philosophical Journal*, Vol. 3, No. 6, 1820, pp. 367–80

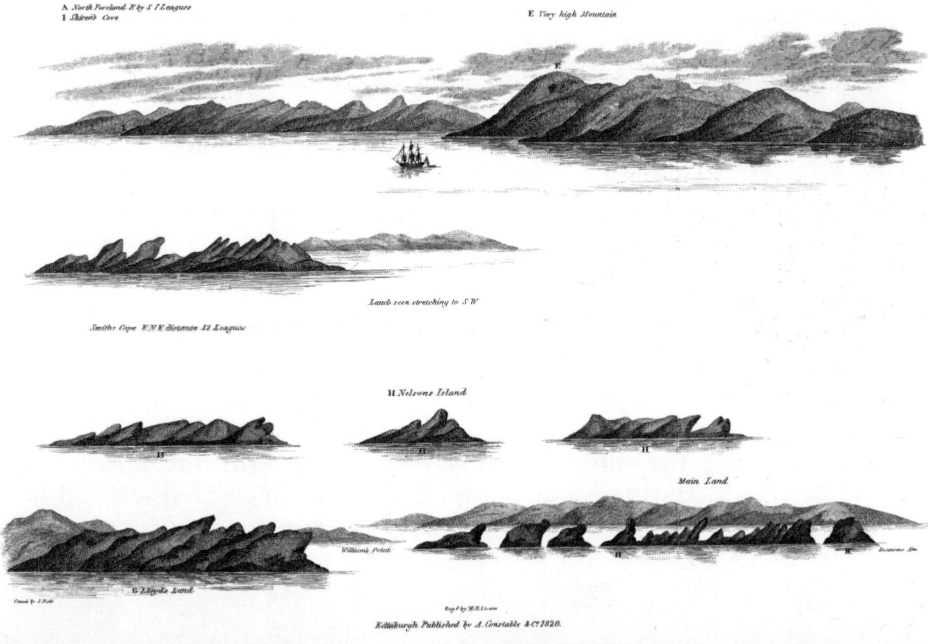

James Weddell

Nathaniel Palmer

By the early 1830s sealing in the South Orkney and South Shetland Islands was no longer profitable and the sealers were once again obliged to extend their search elsewhere. But not all sealers were ruthless profiteers; some, like Weddell, were keen observers of nature and interested in discovery for its own sake. British interest in the Antarctic owes more than is generally recognized to the enterprise of a few public-spirited London oil merchants best exemplified by the firm of Samuel Enderby and Sons. This old-established company was the first to send a British whaling vessel south of Cape Horn into the Pacific Ocean in 1775 and thereby open up the British Southern Whale Fishery. Their early efforts to develop this trade, largely in the hands of the Americans before the War of Independence, were partly frustrated by the opposition of the East India Company who had a commercial monopoly in these southern waters and wished to confine as much trade as possible to their own vessels. It was the Enderbys, heading the other London oil merchants, who after years of lobbying and petitioning succeeded at last in breaking the East India Company's monopoly and establishing the principle of free trade in the southern oceans. The business enterprise of the Enderbys was matched by their interest in both exploration and scientific discovery to which they were always ready to sacrifice commercial advantage and profit. Charles Enderby, a founder-member of the Royal Geographical Society, took the unprecedented step of breaking with the traditional secrecy associated with sealing firms by depositing many of his captain's logs in the Society's library and publishing important extracts in their *Transactions*. By far the most significant of these logs is that of John Biscoe, a veteran of the Napoleonic Wars, who between 1831 and 1832, in the Enderbys' service, circumnavigated the entire Antarctic continent in two tiny vessels, the *Tula* and *Lively*. During his first cruise in 1831 he discovered land in the Australian sector of the continent which he named, after his owners,

'Enderby Land'. The following year he made further new finds along the west coast of the Antarctic Peninsula – Adelaide Island, named after the consort of William IV, and Graham Land after the then First Lord of the Admiralty, Sir James Graham. Notwithstanding the considerable financial loss incurred by this voyage Charles Enderby's ambition to promote further Antarctic discoveries was undiminished. In 1833 one of his captains, Peter Kemp in the *Magnet*, discovered Heard Island and Kemp Land adjoining Enderby Land. In 1839 John Balleny, with the *Eliza Scott* and *Sabrina*, discovered the Balleny Islands and Sabrina Coast, now part of Australian Antarctic Territory.

By the opening years of Queen Victoria's reign, after over half a century of sporadic exploration, the limits of Antarctica were beginning to emerge on the charts. As to whether it was a great southern land or possibly a vast mass of islands only further extensive investigation could determine. Private enterprise, on the lines of the Enderby voyages, could no longer support either the scale or the cost of such undertakings; the next great step forward was to be made at national level.

D'Urville, Wilkes and Ross

Between 1837 and 1839 three nationally organized expeditions entered the field, one French, one American and one British. The motivating force behind the French and British undertakings was the discovery of the South Magnetic Pole. Magnetic research at this time made a claim on international science similar to world weather observations today. The development of steam propulsion and the gradual replacement of wooden vessels with ships of iron gave rise to numerous problems relating to navigation

Dumont d'Urville

Discovery of Terre Adélie by Dumont d'Urville, 20 January 1840

by compass, previously of no great consequence. An accurate knowledge of the Earth's magnetic strength and direction at any part of its surface was becoming essential. The great German physicist, Karl Friedrich Gauss, had predicted that the South Magnetic Pole would be found in the region of latitude 66°S., longitude 146°W. Such an objective was close to the heart of the leader of the French expedition, Jules-Sébastien-César Dumont d'Urville, less well known to posterity as the discoverer of the Venus de Milo. His patron, King Louis-Philippe, was more interested in promoting the glory of France than forwarding the cause of science and d'Urville's two ships, the *Astrolabe* and the *Zelée*, were ill-fitted to enter the heavy pack of the Southern Ocean. They failed, therefore, in one of their principal objectives, the penetration of the Weddell Sea. But in the region of the Antarctic Peninsula a number of islands, already known to the sealers, were mapped and named. Yet more significant was Dumont d'Urville's discovery, in January 1840, of continental land between latitudes 120° and 160°E. in the region of the magnetic pole. This ice-bound coast he named after his wife Adélie; today, Terre Adélie, as it is officially termed, is the chief centre of France's scientific activity in Antarctica, and her principal scientific station is appropriately named Base Dumont d'Urville.

It was while d'Urville was pursuing his course towards the magnetic pole that a chance encounter occurred. While hove-to in a fog a strange vessel, an American man-of-war, passed within hailing distance of the French ships making no attempt at communication. Dumont d'Urville, ignoring this apparent snub, proceeded to chart and explore a further stretch of the Antarctic coastline before sailing northwards for home. But what of the mysterious American ship? She was the *Porpoise*, one of five vessels which comprised the United States Exploring Expedition dispatched to the Pacific by the United States Government in 1838. The expedition's aims were to further knowledge of the southern whale fishery and to carry out some scientific exploration. Certainly, Antarctic discovery was by no means its primary function, indeed neither the ships themselves nor those who manned them were suitably equipped for a prolonged season in the pack ice. Only the iron will of the commander, Lieutenant Charles Wilkes, and his determination to promote American interests in this newly emerging region, can account for the remarkable achievements that were in fact made. Like many before and after him, Wilkes failed to penetrate the Weddell Sea, but one of his squadrons, sailing south-westwards, very nearly beat Cook's record of 71° 10′S. It was only by dint of much improvised repair work to his battered ships that Wilkes was able to embark on a second Antarctic season. In 1839, only hours before Dumont d'Urville, he discovered the coast which now bears his name, charting a series of landfalls as far as Knox Coast. Alas for Wilkes's perseverance and his patriotism; his was no hero's welcome on returning to his native land but a court martial arising from charges of alleged harsh treatment of his crew; he was convicted and heavily fined. Though many of Wilkes's landfalls were subsequently challenged as erroneous, he made valuable contributions to knowledge of the Pacific sector of the Antarctic coastline extending the discoveries made by John Balleny in 1839. It was Wilkes, incidentally, who first named this coastline 'the Antarctic continent', an inspired guess which had yet to be demonstrated by facts.

Charles Wilkes

Sir James Clark Ross

The British expedition that sailed down the English Channel in September 1839, bound for the Antarctic, was to prove even more successful than that of the French and Americans. Its aim was the achievement of the South Magnetic Pole; its leader, appropriately enough, was Sir James Clark Ross, who in 1831 had discovered the North Magnetic Pole in the Canadian Arctic. Ross's expedition was in every respect better prepared for work in Antarctic waters than those of Dumont d'Urville and Wilkes. Its aims were clear cut; its crews all picked men and well equipped; its vessels, *Erebus* and *Terror*, with their double decks, double hulls, watertight compartments and reinforcing beams, were the first ships ever to be thus strengthened for working in ice. Such careful preparation, coupled with the determination and enterprise of the commander, required only good fortune to guarantee success. And this too was to be vouchsafed. Sailing from Tasmania, Ross took a more easterly tack than Wilkes and so by chance penetrated the great embayment of what is now called the Ross Sea. Forcing his way through the fringing pack ice Ross found himself in clear water and beheld in the far distance a new land of snow-covered mountains. Sailing past the dark cliffs of Cape Adare he found himself within full sight of those towering peaks which he named the Royal Society and Admiralty ranges. A landing was made on Possession Island and the newly discovered lands claimed for Queen Victoria. This is the region now known as the Ross Dependency and is administered by New Zealand. Ross's good fortune continued to hold; a few days after the landing the flames and smoke of a distant volcano were seen; this Ross named Erebus, and its smaller extinct neighbour, Terror. Sailing westward of the island which the two volcanoes dominate, now named Ross Island, the expedition sailed into McMurdo Sound in the hope of finding open sea to the south and an unimpeded route to the magnetic pole. But progress southwards was at an end; a great wall of ice loomed up before them with dazzling white cliffs 100 ft. high, as impenetrable as the cliffs of Dover and stretching as far as the eye could see. This was the Great Southern Barrier, the edge of the Ross Ice-Shelf which, as we know now, covers an area larger than France.

Ross was to spend two further summer seasons between 1842 and 1843 in Antarctic waters before returning to England. During the first he continued to explore the barrier his ships narrowly escaping crushing by icebergs during a violent storm. During the second Ross discovered and claimed the James Ross Island group off the northern Antarctic Peninsula after failing to enter the Weddell Sea.

The Antarctic discoveries of Sir James Clark Ross were the most important of the nineteenth century; it was from this sector of Antarctica, which he discovered, that the first great land journeys towards the South Pole were to be launched over 50 years later.

After Ross to the first wintering

Despite the lack of enthusiasm on the part of governments for any further Antarctic exploration in the years that followed James Clark Ross's expedition there was no lack of enthusiasts anxious to resume exploration and no lack of theories as to the relationship of the fragments of land so far discovered. As early as 1860 a plan for international co-operation in the scientific investigation of the region was being mooted. As to the vexed problem of whether Antarctica was a continent or a group of islands this was to be resolved, long before the first serious land exploration, as a result of the outstanding oceanographic cruise of H.M.S. *Challenger*. The task of her commander, Captain George Nares, was to take soundings and dredgings over a vast area of the world's oceans. *Challenger* was the first steam vessel to cross the Antarctic Circle in February 1874. Though the expedition never passed within sight of Antarctica itself it brought back rocks dredged up from the neighbouring ocean-bed which on later geological analysis proved to be of undoubted continental origin.

The exploration of Antarctic waters was only a part, and that a minor one, of *Challenger*'s duties; the money and resources necessary for a co-ordinated and sustained scientific effort in the Antarctic were still not forthcoming. In the event Antarctic exploration was revived by the whaling industry. By the end of the nineteenth century, the traditional Arctic whale fishery was almost exhausted. But the steam whaler, armed with the modern harpoon gun, the invention of the Norwegian Svend Foyn, was

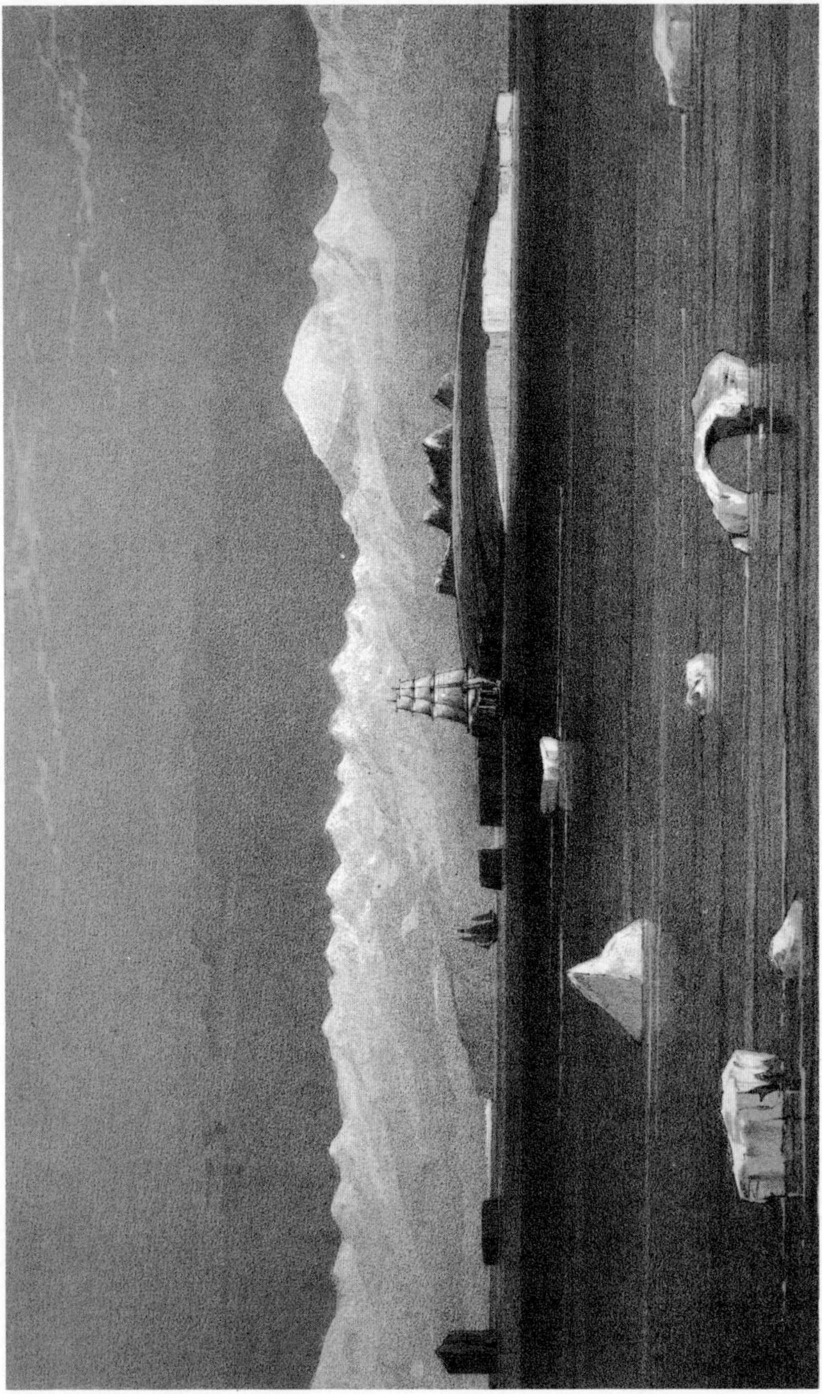

Mount Sabine and Possession Island discovered 11 January 1841. Reproduced from Sir J. C. Ross: *Voyage . . . in the southern and Antarctic regions . . .* 1847

to make whaling in the Southern Ocean an economic proposition. Two such expeditions left for the Antarctic in 1892, the first a Scottish undertaking from Dundee. This pioneer British whaling voyage achieved little but is remembered for one of its naturalists, William Spiers Bruce, who was later to command his own expedition to the Antarctic. The second expedition, led by a Norwegian whaler, Captain Carl Larsen, made important discoveries along the Weddell Sea coast of the Antarctic Peninsula, as far south as latitude 64° 40′S. Larsen, on his return to Norway, reported his successes to Svend Foyn who, in turn, gave financial support to a private expedition led by Henrik J. Bull in the whaler *Antarctic*. Bull, who sailed south in 1894, caught few whales but had the distinction of being first to land a party on the main Antarctic continent, on 24 January 1895, in the region of Cape Adare, Victoria Land.

In Europe a wave of enthusiasm for Antarctic exploration reached its climax in 1895 with the Sixth International Geographical Congress in London which resolved that 'the exploration of the Antarctic regions is the greatest piece of geographical exploration still to be undertaken' and recommended scientific societies throughout the world to 'urge in whatever way seems to them most effective that this work should be undertaken before the close of the century'. There was no immediate response, but three years later two private expeditions entered the field. In 1898 a Belgian expedition, led by Adrien de Gerlache de Gomery, made some new discoveries in the region of the Palmer Archipelago and sailed as far south as Alexander Island. The expedition's ship, the *Belgica*, was then beset in the pack ice and drifted for 12 months to the south of Peter I Øy thereby gaining the distinction of being the first exploring vessel to winter in the Antarctic. This harrowing experience is well recounted by an American member of the expedition, Frederick A. Cook, in his book *Through the First Antarctic Night*. Cook was later to achieve some notoriety as a rival to Peary for the achievement of the North Pole. Another member of this very cosmopolitan expedition was Roald Amundsen, soon to achieve fame as conqueror of the South Pole. The second expedition was nominally a British one sponsored by Sir George Newnes, pioneer of modern popular journalism and the tabloid Press. Its leader was a young Norwegian, Carsten Borchgrevink, who had sailed before the mast on Bull's *Antarctic* in 1894–5. Borchgrevink was more fortunate than the Belgians. His ship, the *Southern Cross*, succeeded in reaching Victoria Land in the region of Cape Adare. Here the expedition wintered, the very first to do so on the Antarctic continent. Some scientific work was achieved including a sledge journey over the Ross Ice-Shelf using sledge dogs in charge of two Lapps.

The twentieth century dawned with the Antarctic continent still largely 'Terra Incognita' after 125 years of spasmodic and ill-co-ordinated exploration. But change was in the air; whereas scientific research had in the past to play second fiddle to the needs of geographical discovery or commercial

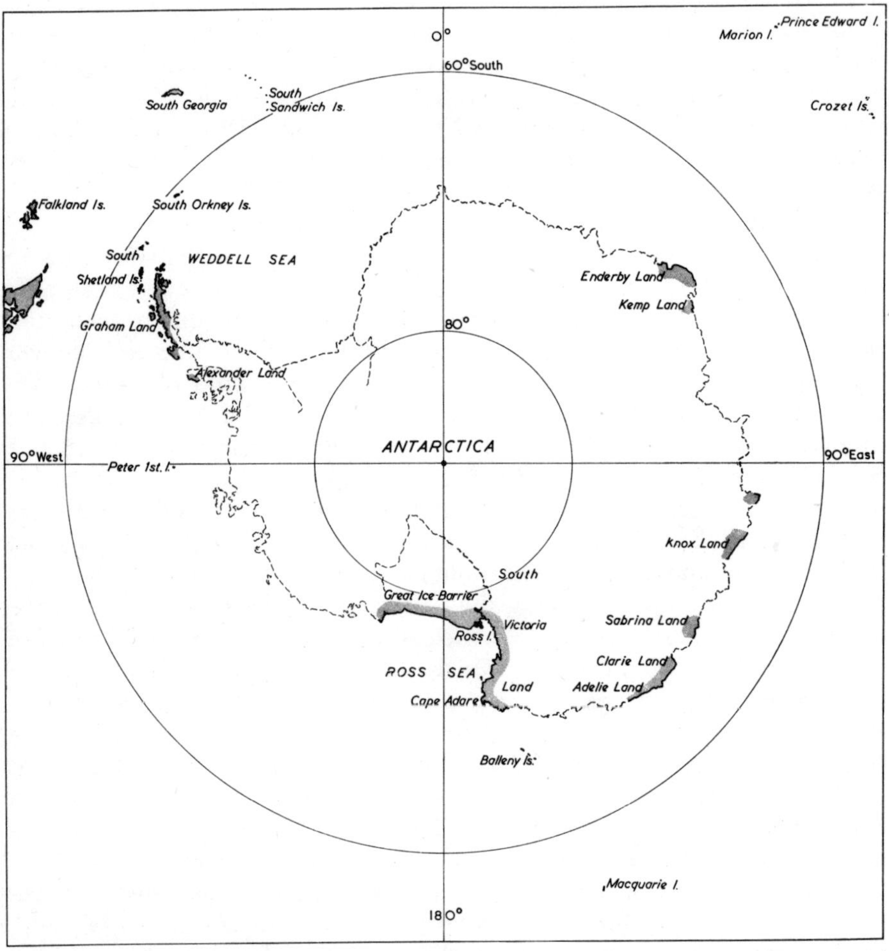

The Antarctic in 1901 showing its largely unknown nature. Adapted from a map published by the Royal Geographical Society, 7 November 1904 (with contemporary place-names retained)

gain, now a new trend was to be discerned. Soon no Antarctic expedition could hope for support, whether public or private, without an approved scientific programme, though it would be many years before such expeditions would be led by trained scientists. The years before the First World War were to witnesss several important expeditions of this kind; some are enshrined in the history books, others are scarcely known to the general public.

Scott's first expedition: the beginnings of the heroic era

To most Britishers the word 'Antarctic' is synonymous with the name of Robert Falcon Scott. At the turn of this century Scott, then a young lieutenant in the Royal Navy, had never thought of himself as a polar explorer. Opportunity knocked when, as a result of a purely chance encounter with Clements Markham, President of the Royal Geographical Society, Scott was encouraged to apply for the post of leader of an expedition about to sail for Antarctica under the auspices of the Royal Geographical Society and the Royal Society of London. Somewhat to his surprise, Scott was appointed. The National Antarctic Expedition, as it was officially called, was predominantly naval in character; its officers were selected from both the Royal and Merchant Navies. One young Merchant Navy sub-lieutenant was Ernest Shackleton, later to achieve fame in his own right as an Antarctic explorer, rivalling Scott as a public idol. The story of this, the first extensive scientific expedition on the mainland of Antarctica should be read in Scott's own *Voyage of the Discovery* and the more recently published diary of Edward Wilson, *Diary of the 'Discovery' Expedition*. Something of the atmosphere of these relatively small-scale expeditions can be gained by visiting *Discovery* herself at her moorings off London's Thames Embankment. Like Nansen's *Fram* she was designed expressly for navigation in heavy pack ice. The *Discovery* expedition achieved the first extensive exploration on land in Antarctica. From their base on Ross Island, McMurdo Sound, Scott and his men accomplished two summer seasons of extensive sledging achieving a record of latitude 82° 17′S., and ascending the continental ice-sheet. And sheer geographical discovery apart much useful work in the fields of meteorology and natural history was accomplished.

Other Antarctic expeditions of this period, several of which made important contributions to our knowledge of the continent, are perhaps less well known. A Swedish expedition, led by Otto Nordenskjöld, sledged as far south as 66° on the east side of Graham Land. The expedition's ship *Antarctic* was crushed in the ice of the Weddell Sea and its crew obliged to spend the winter in three separate parties until rescued by the Argentine vessel *Uruguay*. France's present interest in Antarctic exploration stems from the expeditions of Jean-Baptiste Charcot who sailed twice to the peninsula region of Antarctica before the First World War; in 1903–5 in

The kitchen of Scott's hut at Cape Evans, Ross Island, 1910–13, restored by the New Zealand government over half a century later

Scott's hut at Cape Evans as it looks today

Members of Scott's last expedition. *Left to right:* Oates, Lieutenant Evans, Bowers, Wilson, Scott

Robert Falcon Scott

Wilson, Bowers, Seaman Evans, Scott and Oates at the South Pole, 18 January 1912

Otto Nordenskjöld

Jean-Baptiste Charcot

the *Français* and in 1908–10 in the *Pourquoi Pas?*. Though lacking in the drama that makes headlines Charcot's expeditions made numerous discoveries off the west coast of the peninsula and carried out a great deal of careful charting. Contemporaneous with Scott's *Discovery* Expedition was a German expedition, under Erich von Drygalski, which in 1902–3 first discovered Wilhelm II Land in present Australian Antarctic Territory, and whose steam vessel, *Gauss*, was at one point compelled to stoke its furnace with penguins. William Spiers Bruce's Scottish National Antarctic Expedition of 1902–4 carried out the first oceanographic exploration of the Weddell Sea, discovered Coats Land and set up a meteorological observatory on Laurie Island in the South Orkney Islands, still operated today by the Argentine Meteorological Service.

Such expeditions added greatly to our knowledge of the Antarctic regions but they never gripped the popular imagination as did the race for the North and South poles. The attainment of these purely artificial concepts seemed as much a patriotic duty at that time as a moon landing does today. The part played now by television and radio in building up an astronaut's 'image' and bringing his achievements into the homes of the people was accomplished for the polar heroes by the popular newspaper and magazine Press served by the newly invented wireless telegraphy. No ambitious explorer, however much he might privately despise such publicity, could afford to ignore 'news value' in his plans; Antarctic exploration is an expensive undertaking and government support was not so readily forthcoming then as it is today.

The siege of the South Pole

No polar explorer courted publicity more effectively than Ernest Shackleton. Invalided out of Scott's *Discovery* Expedition by scurvy, a nutritional disease whose cause was to defeat medical science until after the First World War, Shackleton determined to prove himself by achieving the South Pole, much as Robert Peary, the American explorer, was trying for the North Pole. Sailing in the *Nimrod* in 1907 Shackleton wintered at Cape Royds, on Ross Island, and succeeded in sledging to within 97 miles of his heart's desire by way of a glacier, newly discovered, which he named after his patron, the industrialist William Beardmore. A second party, under the Australian Edgeworth David, succeeded in reaching the vicinity of the South Magnetic Pole, then in latitude 72° 25′S., longitude 155° 16′E. In April 1909, at the other end of the world, Peary finally reached the North Pole, leaving the South Pole as the Earth's last great geographical prize. The eyes of the world watched to see who would claim it. Even before Shackleton had gone south, Scott had planned to make a return visit. Shackleton's near success clinched the idea in his mind and determined Scott to achieve the South Pole. So it was that the British Antarctic Expedition, with a comprehensive scientific programme and the South Pole as its declared objective, left Cardiff Docks in June 1910 for Ross Island by way of Australia and New Zealand. On arriving at Melbourne Scott received a telegram the contents of which cast a shadow across the whole enterprise: 'Beg leave to inform you proceeding Antarctica, Amundsen.' Cheated of the North Pole by Peary, the Norwegian, determined not to be forestalled in the Antarctic by Scott, had made a sudden and, until then, unannounced change of plan. Inevitably what should have been an unhurried and largely scientific venture became a race. The story with its tragic sequel needs no repetition here. There are no substitutes for Scott's own published diary, *Scott's Last Expedition*. The story is also told in Cherry-Garrard's *The Worst Journey in the World*. Apsley Cherry-Garrard, youngest of Scott's scientific staff, accompanied Wilson and Bowers on the hair-raising midwinter journey to Cape Crozier to collect the eggs of the Emperor Penguin. Though he did not accompany Scott all the way to the Pole, to Cherry-Garrard fell the hopeless task of attempting the relief of Scott and his companions in March 1912. The following November he had the agonizing experience of accompanying the search-party which eventually found the tent wherein lay, embalmed by the cold, the bodies of Scott, Wilson and Bowers, together with the precious diaries telling of the events at the Pole and of the deaths of Evans and Oates. Much paper and print has been devoted to arguing the reasons for Scott's failure. Enough here to say that Amundsen achieved his goal by abandoning all pretence at science and skilfully employing his dogs in the attainment of a fixed objective. Scott had no great faith in dogs and preferred to rely on man-power for the last leg of the pole journey. Nor was he inclined to

Sir Ernest Shackleton

Shackleton's ship *Endurance*. She was crushed by the Weddell Sea pack ice and sank on 21 November 1915

Shackleton's hut at Cape Royds, exterior

To earn money for his expeditions, Shackleton devoted much time and enormous energy to lecturing. He lectured at the Philharmonic Hall, London, six days a week for five months in 1919–20

PLEASE TAKE THIS WITH YOU.

PHILHARMONIC HALL

GT. PORTLAND STREET, W

(Near Oxford Circus),

By arrangement with Mr. ANDRE CHARLOT

TWICE DAILY

2.30 and 8.30

The EVENING STANDARD says:
Both the lecturer and the pictures provide one of the finest entertainments that have ever been known in London.

SIR ERNEST

SHACKLETON

HIMSELF shows the

Marvellous Moving Pictures

and tells the story

OF HIS LATEST ANTARCTIC EXPEDITION

The Times says—"This entertainment has become one of the most popular in London."
The Daily Telegraph says—"Everybody who goes will have a good time."
The Daily Chronicle says—"Those who miss it will miss one of the finest things which London can give us at the present day."
The Daily Mail says—"Easily the best picture show in London."
The Westminster Gazette says—"A remarkable entertainment which all London should see."
The Daily Express says—"Shackleton held the audience spellbound."
The Daily Mirror says—"The film is among the wonders of the world."
The Evening News says—"One of the best entertainments in London."

Reserved **7**s. **6**d. **5**s. and **4**s. Unreserved **3**s., **2**s. and **1**s. (Tax Extra)
From the BOX OFFICE at the Hall (Telephone: Mayfair 3003) and the usual Libraries

jettison entirely his scientific programme, a positive disadvantage in the circumstances. Unexpectedly bad weather, a late start, shortage of fuel and food leading to premature exhaustion were all factors contributing to the disaster. Yet in a sense Scott's failure was a victory of another kind; his diary and his last letters have proved an inspiration and a lesson to future generations in the art of living and dying.

Shackleton's *Endurance* Expedition: the end of the heroic era

The achievement of the South Pole did not conclude the so-called 'heroic era' of Antarctic exploration. There remained a further goal, perhaps even more ambitious, the crossing of Antarctica. This would require a two-pronged attack, one from the Weddell Sea the other from the Ross Sea coasts. Such a plan had already been advocated by the Scotsman, W. S. Bruce, and by a German, Wilhelm Filchner. Both were anxious to prove, or disprove, a long-entertained notion that there might be a channel dividing Antarctica between the Ross and Weddell Seas. Filchner, unable to find the resources for the enterprise, contented himself with an expedition in the ship *Deutschland* to the Weddell Sea coast where his name is commemorated today by the Filchner Ice-Shelf. But to a born gambler like Shackleton the idea of a trans-Antarctic crossing was irresistible. In 1914, as the war clouds gathered over Europe, Shackleton, with the blessing of Winston Churchill, sailed on board *Endurance* bound for the Weddell Sea. His plan was to land in Bruce's Coats Land and eventually meet up with an independent Ross Sea shore-party which would lay advance depots across the ice-shelf up to the Beardmore Glacier. But *Endurance* never reached Antarctica; she was beset in the pack ice and slowly crushed. Shackleton and his crew were forced to camp on the floating ice. Miraculously they were borne by the currents towards Elephant Island in the South Shetland group. Leaving the main body of the expedition in the relative safety of this island Shackleton, with a small picked crew of five, sailed in one of the ship's boats, the *James Caird*, across the storm-tossed Southern Ocean to South Georgia there to seek help from the whaling station. The story of this perilous undertaking and the subsequent trek across the mountainous interior of South Georgia is one of history's epics. After several abortive attempts at rescuing Shackleton's men, success was finally achieved and the Elephant Island party taken off by the Chilean vessel *Yelcho* in August 1916. Shackleton's expedition added little to the geographical knowledge of Antarctica but it proved to be the inspiration behind another far better equipped enterprise of 40 years later; the Commonwealth Trans-Antarctic Expedition of 1955–8.

Antarctic exploration between the two world wars

The theme of Antarctic exploration between the world wars is one of great acceleration in the rate of geographical discovery. In 1920 over half the coastline of Antarctica remained uncharted and the interior, with the exception of the narrow corridor from the Ross Ice-Shelf to the Pole, largely a blank. By 1940 much of the coastline was on the map and the less accessible regions of the continent probed in several places. The reasons for this great leap forward were primarily technical, especially remarkable being the development of light aircraft for surveying. Motives, as always, were mixed; basically they were a combination of commercial and political interests from which science was to reap many advantages. The work of the British Discovery Committee exemplifies this pattern at its best. Established by the Colonial Office in 1923 the Committee's brief was to consider the best means of preserving a then flourishing whaling industry and to promote scientific investigations in the Antarctic. Between 1925 and 1939 some 13 separate cruises were made in southern waters by *Discovery* (Scott's old ship) and other vessels. These Discovery investigations, as they came to be called, were mainly concerned with oceanography and marine biology; they have justly been described as the first real effort at sustained research in the Antarctic.

The establishment of the Discovery Committee followed logically from Great Britain's claim, in 1917, to the regions of the Antarctic mainland and the adjacent islands thereafter to be known as the Falkland Islands Dependencies. To substantiate a territorial claim in international law it is necessary not merely to add the newly discovered land to the map but to occupy and exploit it. This the British Government was determined to do in the Antarctic using the income from the leasing of whaling rights to finance research on the whales themselves. Other nations were quick to establish their claims; New Zealand in 1923, France in 1924, Argentina in 1925 and Australia in 1931. Norway, too, on the basis of a series of expeditions between 1927 and 1937 asserted her sovereignty over a section of Antarctica in 1939. United States' explorers, such as Admiral Richard E. Byrd and Lincoln Ellsworth, also made claims on behalf of their government; these claims have in fact never been taken up though the United States has always reserved the right to do so.

So much for the general trend of Antarctic exploration between the two world wars. New discoveries were to be made not only in such relatively well-known regions as the Antarctic Peninsula but also along the vast stretches of unknown coastline representing the sectors of Antarctica now known as Byrd Land, Dronning Maud Land and Australian Antarctic Territory.

The advent of aircraft

In the peninsula history was made in 1928 with Sir Hubert Wilkins's introduction of the first aircraft to the Antarctic in a bid to fly across the continent from the Weddell to the Ross Seas. Both attempts, in 1928 and

Single-engined De Havilland Fox Moth used on the British Graham Land Expedition, 1934–7

1930, were failures but Wilkins returned with some remarkable air photographs which seemed to show that the peninsula was separated from the continent by a series of channels. The trans-Antarctic flight was eventually achieved by the United States' explorer Lincoln Ellsworth who in 1935, with Hollick Kenyon as pilot, flew from Dundee Island, in the South Shetland group, to the Bay of Whales on the Ross Ice-Shelf. One of the most scientifically productive expeditions of this period was John Rymill's British Graham Land Expedition of 1934–7 which sledged down the west coast of the peninsula, discovered George VI Sound and, by travelling eastwards across the base of the peninsula, was able to show that Wilkins's channels were non-existent.

The Americans return to the Antarctic

While this activity was taking place in the peninsula its hitherto unexplored continuation southwards towards the Ross Sea was being investigated by one of the most remarkable of polar explorers, the American Richard Evelyn Byrd. Fresh from conquering the North Pole by air in 1926 Byrd, advised by Roald Amundsen, turned his attention to the Earth's southern hub. From Little America I, his base on the Ross Ice-Shelf, Byrd successfully flew across the South Pole in November 1929. He also discovered large areas of Edward VII Land and the region to the west of the Ross Ice-Shelf which he named after his wife Marie Byrd (now Byrd Land). In recognition of his work the United States Government advanced him to the rank of Rear-Admiral by a special Act of Congress. A second

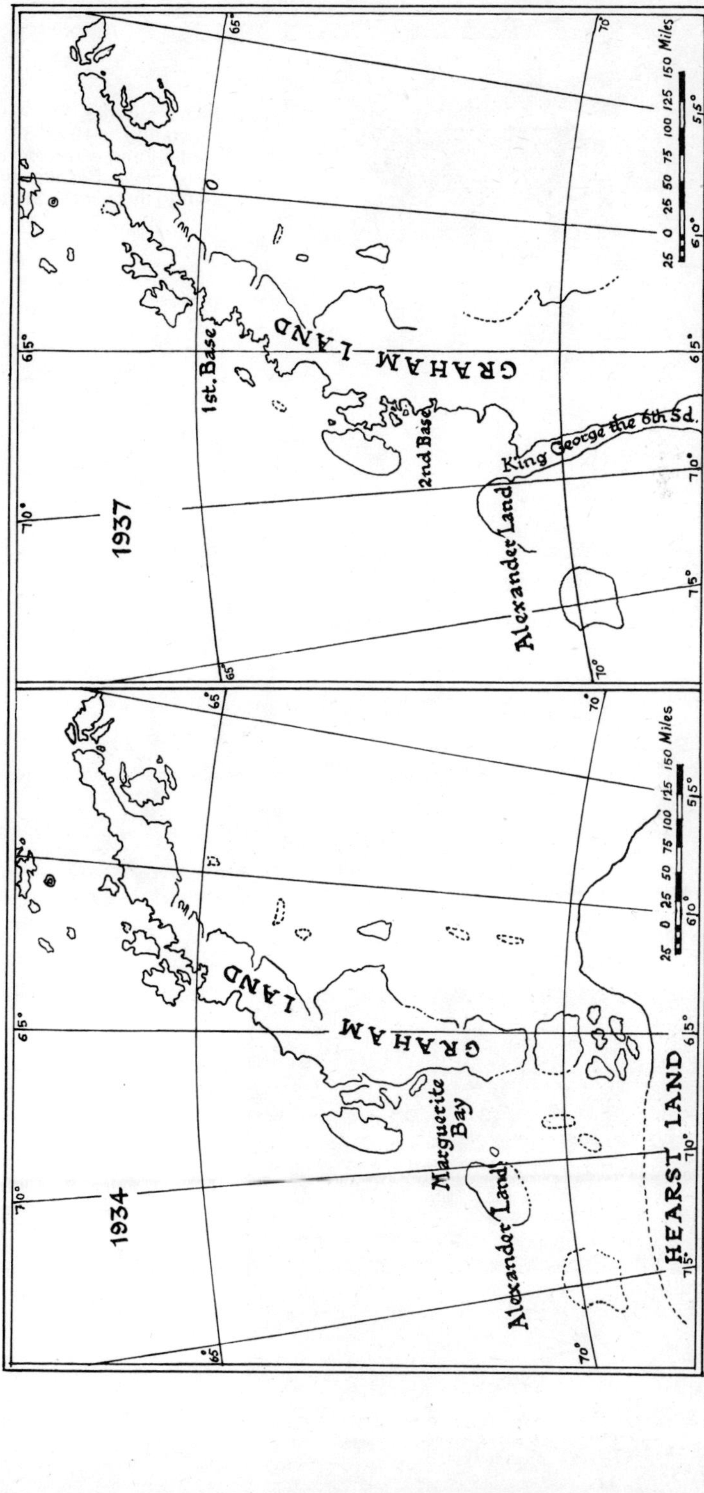

Map of the Antarctic Peninsula before and after the British Graham Land Expedition, 1934–7

Richard Byrd's first Antarctic expedition, 1928–30. The Ford three-engined monoplane *Floyd Bennett*, first aircraft to fly over the South Pole

Byrd's flagship *City of New York*

Little America, Byrd's base on the Ross Ice-Shelf, during the Antarctic night. This was the first Antarctic expedition to maintain regular two-way radio communication with civilization

expedition in 1933–5 was equally successful and Byrd, by means of further air photography, was able to show that the long-conjectured channel linking the Ross and Weddell Seas was a myth. Admiral Byrd was not only the first to envisage the use of aircraft in the reconnaissance of remote and inaccessible regions but was also a pioneer of motor transport and radio communications for field-parties. It was Byrd's first two expeditions that were largely instrumental in drawing the attention of the United States Government to the Antarctic for the first time since Charles Wilkes's expedition of 1838–42. In 1939 the United States decided to lay the basis of a permanent occupation of Byrd Land by sending an expedition south with Byrd himself again in command. Two bases were established on this occasion, one on Stonington Island, off the west coast of the peninsula (East Base) and a second (West Base) on the Ross Ice-Shelf at Little America III. Extensive exploration was carried out in both sectors but the United States declaration of war on Germany brought the expedition to a premature close.

Norwegian and German discoveries

The exploration of the Antarctic coastline between the Weddell Sea and Enderby Land was largely the work of the Norwegians at the initiative of the Christensen family of Sandefjord who successfully reconciled the demands of a profitable whaling business with the less obviously remunerative pursuit of Antarctic exploration. In the course of a whole series of expeditions between 1928 and 1938 Christensen ships, with the aid of aircraft, explored the coasts of Enderby Land and the neighbouring sector which they named for Queen Maud of Norway, Dronning Maud Land. To this same part of Antarctica came an expedition from Hitler's Germany under the patronage of Hermann Göring; after completing a subsequently invaluable air survey the area was bombed with metal swastikas and claimed for the Reich as 'Neu-Schwabenland'.

The last remaining stretch of Antarctica to be explored was that lying between Wilhelm II Land, discovered by Drygalski's German expedition in 1901–3, and Enderby Land, sighted by John Biscoe in 1831. Exploration of this area was largely the work of a joint British-Australian-New Zealand Expedition (BANZARE) between 1929 and 1931 under the leadership of the veteran Australian explorer Sir Douglas Mawson. Both Princess Elizabeth Land and Mac.Robertson Land were added to what, in 1932, was to be named Australian Antarctic Territory. At the same time the boundary between Australian-claimed Enderby Land and Norwegian-claimed Dronning Maud Land was amicably fixed at longitude 45°E.

Exploration after the Second World War

The pattern of national rivalry which had characterized Antarctic exploration between the wars continued both during the Second World

Lars Christensen, head of the Norwegian whaling firm whose ships explored the coasts of Antarctica between 1928 and 1938

Sir Douglas Mawson

War and in the years that followed. In the Antarctic Peninsula, where Argentina had been joined in 1940 by Chile as a rival to Britain's Antarctic claims, there was to be constant quarrelling and counterclaims to sovereignty. Fortunately the shots fired in anger were few. But from the evils of necessity much good was to spring. In 1943 Britain sent a naval expedition, code-named Operation Tabarin, to establish permanent bases in the South Orkney and South Shetland Islands in a bid to reassert her sovereignty. In 1944 a permanent programme of scientific research was established whose administration was transferred a year later to the Colonial Office under the name of the Falkland Islands Dependencies Survey. Today, as the British Antarctic Survey, it is the organization chiefly responsible for manning British bases in the Antarctic and supervising their scientific work.

Other nations were not slow in following Britain's lead. In 1947 both Argentina and Chile established permanent meteorological stations in the disputed area of the Antarctic Peninsula, variously known as Graham Land, Palmer Land, Tierra San Martín and Tierra O'Higgins according to national preference. In the same year Australia created the Australian National Antarctic Research Expeditions (ANARE) which was henceforth to operate permanent Antarctic stations and carry out a comprehensive scientific programme. France, too, began to send regular expeditions to her most southerly territories and in 1950 built a permanent research station in Terre Adélie named Port-Martin. The U.S.A., though still making no formal territorial claims, hastened to reserve its historic rights in Byrd Land by launching, in 1946–7, Operation Highjump with Admiral Byrd himself as leader. Highjump was by far the most ambitious

Byrd revisits his old hut at Little America during Operation High-jump, 1947–8

Unloading supplies on to the ice of the Bay of Whales, Ross Sea, during Operation High-jump

John Giæver, leader of the Norwegian-British-Swedish Expedition, 1949–52

exploratory venture yet to be attempted by a single nation in Antarctica involving 13 ships and 4000 men. Though primarily a political and military exercise Highjump did, nevertheless, establish the scale and pattern of the future series of expeditions known as Operation Deep Freeze which brought the U.S.A. permanently to Antarctica in 1954.

With the gradual establishment of permanent bases round Antarctica, the old pattern of spasmodic, unco-ordinated expeditions, frequently forced to rely on the efforts of a few enthusiasts supported by private funds, began to change. Antarctica was now to become the permanent responsibility of nations. And from national responsibility the concept of international responsibility was shortly to develop. A foretaste of this new spirit was provided by the Norwegian-British-Swedish Expedition of 1949–52, the first truly international expedition to the Antarctic. Conceived by the Swedish glaciologist, Professor H. W. Ahlmann, directed by the late Dr. H. U. Sverdrup, then Director of the Norwegian Polar Institute, and led in the field by another Norwegian with much polar experience, Captain John Giæver, it demonstrated that at the scientific level the nations find little difficulty in resolving their problems. The scene of operations was Norwegian Dronning Maud Land where the German 'Neu-Schwabenland' Expedition had carried out a large-scale air reconnaissance in 1938–9. Here representatives of all three nations co-operated in a number of scientific programmes including the first seismic traverse of the inland ice-sheet. Judged purely on its scientific results the Norwegian-British-Swedish Expedition was certainly the most productive since the days of Captain Scott.

Though international co-operation in Antarctica was evidently desirable if science was to make any real progress the prospect of achieving it in the early 1950s seemed remote. The 'cold war' was at its height and the Soviet Union remained aloof and disinterested. By the end of the decade a miracle had taken place. The political stalemate in Antarctica had been temporarily resolved and all claims shelved for a 30-year period. Over a dozen nations, including the U.S.S.R., were amicably co-operating in a mutually agreed programme of scientific exploration of the continent which continues to the present day. The significance of this revolutionary change of attitude was to have tremendous consequences for the future of scientific exploration in the Antarctic. We devote our next, and concluding, chapter to the story of how it came about.

Members of the Norwegian-British-Swedish Expedition burying a seismometer during the first full-scale seismic traverse of the Antarctic ice-sheet

14
THE ANTARCTIC AND THE INTERNATIONAL GEOPHYSICAL YEAR

In a world divided by warring ideologies and soured by international mistrust and misunderstanding it is some comfort to find one area at least of man's activities where reason tends to prevail – Antarctica. Here a dozen nations and more, of many political persuasions and differing tongues, find themselves able to put aside their customary allegiances and work together in the commonwealth of science. The foundations of this utopian state were laid during a period of scientific co-operation which has been fairly epitomized as 'The single most significant peaceful activity of mankind since the Renaissance and the Copernican Revolution'. The International Geophysical Year (IGY for short), as this period of co-operation was known, lasted from 1 July 1957 to 31 December 1958. It involved at its peak over 30,000 scientists and technicians from 66 nations manning more than 1000 stations, not merely in the Antarctic but girdling the whole Earth. Its main concern was with the geophysical sciences, that is, the disciplines concerned with the study of the Earth and the forces which affect it – meteorology and upper air physics, glaciology and oceanography. Antarctica has unique qualifications as a natural laboratory for the study of geophysical operations of all kinds. At the South Geomagnetic Pole, as at the North, the Earth's geomagnetic field is strongest and its effect on various atmospheric phenomena can best be studied here. The ice-sheet of Antarctica greatly influences the climate of the whole Southern Hemisphere; to appreciate Antarctica's function as one of the world's two great heat sinks it is necessary to understand the atmospheric circulation and temperature distribution over the continent. Again, with 90 per cent of the world's ice locked up in Antarctica's ice-sheet a unique opportunity is presented to glaciologists to interpret the history of ice-sheets past and present and to draw conclusions as to the variations of the Earth's climate throughout the ages. The IGY gave scientists an opportunity to carry out these and other studies and to accumulate the exchange of synoptic data. The results were so satisfactory that at the conclusion of the agreed period the nations concerned resolved to prolong international research in the Antarctic indefinitely, adding to the programme of geophysical studies disciplines such as biology, geology and cartography. In 1959 the future of Antarctica as a continent for science was underwritten by an international treaty, the first of its kind in world history.

In this chapter we trace the pattern of events leading up to the IGY and show how 12 nations, abandoning traditional prejudices and attitudes, successfully solved one of the most complex logistical problems of all time, the setting up and the manning of many new stations and observatories on the coast and on the ice-sheet of Antarctica.

The beginnings of scientific co-operation

Co-operation in the exchange of scientific data between the nations is a comparatively recent phenomenon in history. A century and a half ago it was still possible for an individual, blessed with the essential genius, to encompass virtually the whole field of scientific knowledge as it then existed. Such a man was the German naturalist Alexander Baron von Humboldt who was largely responsible for laying the foundations of modern physical geography and meteorology. Humboldt was possibly the first person to secure the scientific co-operation of nations by persuading the Russian and British Governments to establish stations throughout their dominions for the study of geomagnetism and meteorology. Another such genius, again a German, Karl Friedrich Gauss, founded a magnetic union for the simultaneous observation of magnetic phenomena throughout Europe. We have already seen in our previous chapter how Gauss's prediction of the South Magnetic Pole inspired the Antarctic Expeditions

Baron von Humboldt

of Sir James Clark Ross and Dumont d'Urville. This would have been an ideal opportunity for a sharing of resources and an exchange of data but the time was not yet ripe. Probably the true father of international co-operation was an American, Matthew Fontaine Maury, first Superintendent of the United States Hydrographic Office. After successfully launching a scheme for the recording and pooling of weather and ocean data by the navies of several maritime nations, Maury endeavoured to extend it to the exploration of Antarctica so that sailing directions could be amplified and corrected. The idea was heartily applauded and endorsed by the nations but no action followed. Then in 1875 an Austrian Navy lieutenant, Karl Weyprecht, a seasoned explorer of the Russian Arctic, advocated to a meeting of German scientists that scientific programmes in both polar regions should be co-ordinated and the methods of observation standardized. Weyprecht did not live to see his recommendations put into action but they were nevertheless, warmly received by European politicians and scientists. In 1880 an International Polar Commission was set up to initiate the first International Polar Year of 1882–3, in which 12 nations set up 14 stations in the polar regions to make co-ordinated observations of the Earth's magnetism and climate. Only two of these stations were in the Southern Hemisphere, one on South Georgia and one at Cape Horn. Between this and the next similar experiment 50 years were to elapse. The International Polar Year of 1932–3 was partly an attempt to repair the damage done to the structure of international science by the First World War of 1914–18 and was the inspiration of a German Arctic explorer, Johannes Georgi. Like its predecessor, this Second International Polar Year was mainly an Arctic affair. As before, the programme was geophysical with emphasis on meteorology and the upper layers of the Earth's atmosphere, which were believed to reflect radio waves, both matters of immediate relevance to problems of air navigation and communication. For the first time serious attention was paid to the little-understood aurora. A number of new techniques were introduced which anticipated those used in the IGY – auroral cameras for example – and there were ambitious plans to send rocket-borne instruments into the upper atmosphere, though these never matured. It was unfortunate that this Second Polar Year happened to coincide with a world economic depression so that its scale had to be very much curtailed.

Had the 50-year cycle between Polar Years repeated itself Antarctica today might well remain a relatively unknown continent. But once again individual initiative succeeded in arousing enthusiasm for an all-out onslaught on this continent.

The birth of the International Geophysical Year

In April 1950 the American geophysicist, Dr. James Van Allen, after whom the Van Allen radiation belts in space are named, held an informal

The symbol of the International Geophysical Year shows the Earth partly sunlit, partly in darkness, indicating the influence of the Sun on the Earth. The South Pole is shown to emphasize the special attention given to the Antarctic during the IGY and the satellite in orbit denotes the pioneer work in this field at the time. The symbol was used unframed and also framed in an octagonal border with the International Year in two languages, here French and English and Japanese and English

meeting of fellow scientists in his Washington home. Among the guests were Dr. Sydney Chapman of Great Britain, an expert on aurora who had participated in the Second Polar Year of 1932–3, and Dr. Lloyd Berkner, a veteran of Admiral Byrd's First Antarctic Expedition of 1928–30. It was Berkner who mooted the idea of a Third Polar Year to take advantage of the rapid advances in the geophysical sciences and the techniques for exploring the Earth's surface and upper atmosphere that had been made since 1933. The period 1957–8 was seen as an auspicious one for the proposed Polar Year since it was known that sunspot activity would then be at a maximum, making a valuable contrast with the previous Polar Year when it was at a minimum. Berkner's idea was received so enthusiastically by his colleagues at this historic meeting that it was in due course commended to the International Council of Scientific Unions (ICSU). As its name suggests ICSU is a federation of smaller specialist unions which enables scientists from all over the world to meet and discuss their problems at regular intervals. ICSU welcomed the idea and set up a special committee to initiate advance planning. The Committee, which represented

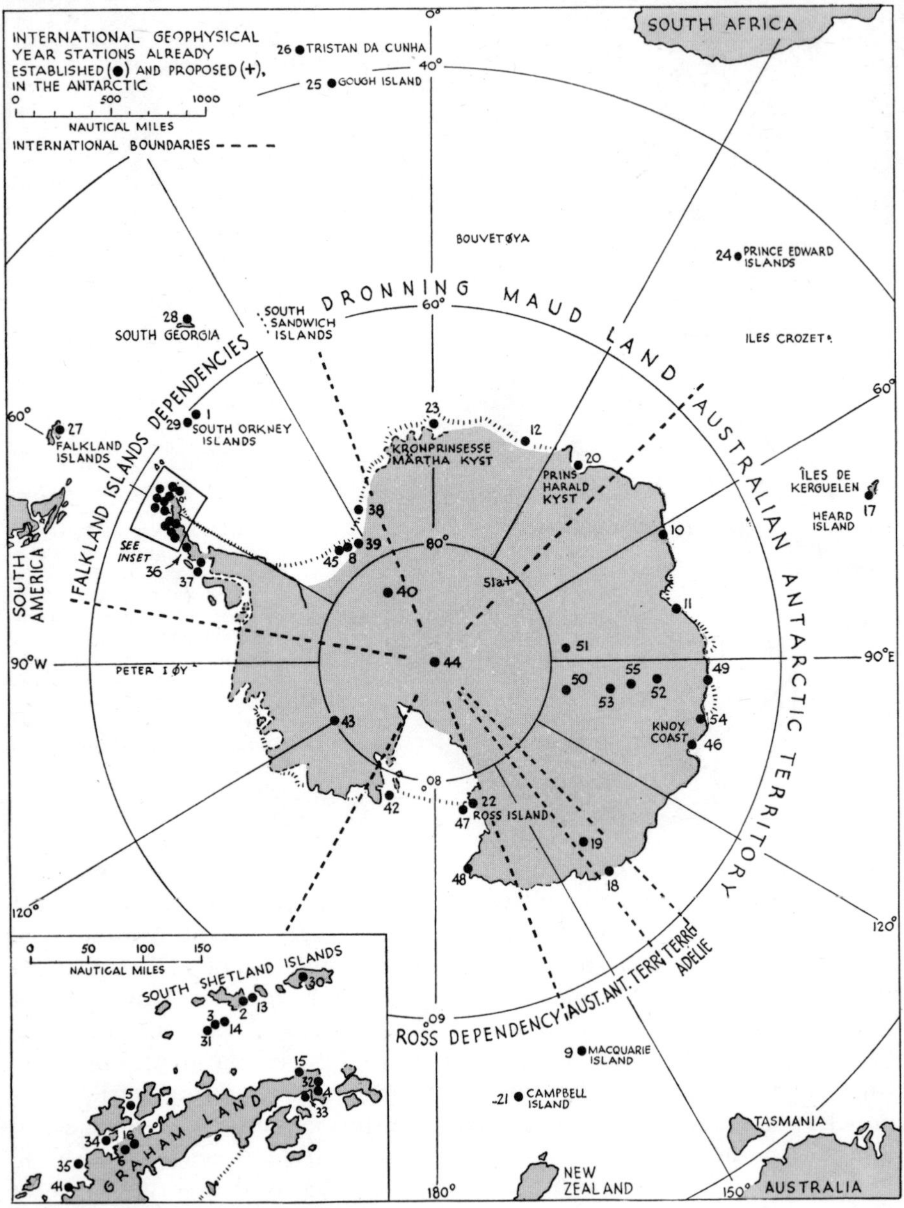

Map showing stations occupied in the Antarctic during the International Geophysical Year, 1957–8. *See key opposite*

ARGENTINA

1 Orcadas
2 Teniente Camara
3 Decepción or Primero de Mayo
4 Esperanza
5 Melchior
6 Almirante Brown
7 General San Martín
8 General Belgrano

AUSTRALIA

9 Macquarie Island
10 Mawson
11 Davis

BELGIUM

12 Roi Baudouin

CHILE

13 Arturo Prat
14 Pedro Aguirre Cerda
15 General Bernardo O'Higgins
16 Presidente Gabriel Gonzalez Videla

FRANCE

17 Port-aux-Français
18 Dumont d'Urville
19 Charcot

JAPAN

20 Syowa

NEW ZEALAND

21 Campbell Island
22 Scott

NORWAY

23 Norway Station

SOUTH AFRICA

24 Marion Island
25 Gough Island
26 Tristan da Cunha

UNITED KINGDOM

27 Stanley
28 Grytviken
29 Signy Island
30 Admiralty Bay
31 Deception Island
32 Hope Bay
33 View Point
34 Port Lockroy
35 Argentine Islands
36 Detaille Island
37 Horseshoe Island
38 Halley Bay
39 Shackleton
40 South Ice
41 Prospect Point

UNITED STATES

42 Little America V
43 Byrd Station
44 Amundsen-Scott Station
45 Ellsworth Station
46 Wilkes Station
47 Williams Air Operations Facility

UNITED STATES/NEW ZEALAND

48 Hallett Station

U.S.S.R.

49 Mirny
50 Vostok
51 Sovetskaya
52 Pionerskaya
53 Komsomol'skaya
54 Oazis
55 Vostok-I

67 interested nations, christened the whole project the International Geophysical Year and itself came to be named the Special Committee for the International Geophysical Year.

By the time the Special Committee had convened a meeting in Rome in 1954 to consider the detailed national programmes, the United States had already drafted one on a massive scale and the Soviet Union likewise came forward with some far-reaching proposals. It was at this Rome meeting that the delegates agreed to concentrate on the two areas which advanced technology had made so much more accessible to man than in the past – outer space and Antarctica; the Antarctic, they felt, was 'a region of almost unparalleled interest in the fields of geophysics and geography alike'. The following year, 1955, a subcommittee met to co-ordinate an Antarctic IGY programme thereby avoiding duplication of effort and deploying the various national resources to the best advantage. There was at one stage a general apprehension that politics might once more rear its ugly head and that the U.S.A. and the U.S.S.R. would clash over the former's dramatic proposal to establish a year-round station at the South Pole itself. But to the relief and gratification of all, the Soviets approved of this ambitious scheme and undertook themselves the equally arduous and novel task of setting up stations at the South Geomagnetic Pole and Pole of Inaccessibility. The spirit of mutual trust and co-operation displayed at this early stage of the IGY has characterized the relationships between all nations active in Antarctica from that day to the present.

Then followed a number of meetings in which the fine details of organization had to be thrashed out to ensure that the scientific observations of the nations co-operating were properly co-ordinated and the final results made available to all who could make use of them. Special days called 'World Days' – three to each month – were fixed for certain activities that could not, for reasons of economy, be carried on all the time – the firing of rockets to explore the upper atmosphere for example. Standard instructions for various disciplines were formulated and published in a series known as the *Annals of the IGY* which were at a later date to include summaries of the results. The all-important problem of international data exchange evoked a resolution from the Special Committee for the IGY that 'all observational data to be exchanged in accordance with the IGY program shall be available to scientists and scientific institutions in all countries'. To house the data itself a number of World Data Centres, or WDCs, were set up in different parts of the world, each of which would make available a complete set of data in all the disciplines studied to scientists in the region it was to serve. Each centre was subdivided into a number of archives dealing with different disciplines, for example glaciology, oceanography, aurora and airglow, rockets and satellites, and so on, each being attached to an institution specializing in the science concerned. Three WDCs were established – WDC 'A' in the U.S.A., WDC 'B' in the U.S.S.R. and WDC 'C' divided among several European countries including Great Britain. An actual example will best illustrate how the system works in practice. At the Scott Polar Research Institute in Cambridge is housed WDC 'C' – glaciology. This archive is based on data relating not only to Antarctica but all the Earth's glaciated regions acquired during the IGY from the many organizations participating. This raw data, that is observations which had not at the time of receipt been analysed and the conclusions published, is supplemented by a considerable library of published books and papers. The WDC makes its data available to all interested scientists for the basic cost of copying and postage. A regular exchange of acquisition lists between the three glaciological centres helps to publicize new material as it becomes available.

In all, 12 nations submitted programmes to the IGY Antarctic Committee. In addition to the two major powers, the U.S.A. and the U.S.S.R., they included Argentina, Australia, Belgium, Chile, France, Great Britain, Japan, New Zealand, Norway and South Africa. Between them these nations were to establish some 60 stations, over 40 of which were on the mainland or on the Antarctic Peninsula; others were sited on various Antarctic and sub-Antarctic islands. The programmes submitted by the 12 nations embraced a wide range of geophysical sciences including meteorology, geomagnetism, upper air studies, glaciology (including seismic and gravity surveys of the Antarctic ice-sheet) and oceanography. Not strictly part of the IGY programme at all was the Commonwealth Trans-Antarctic Expedition, a joint British-New Zealand enterprise which planned to realize Shackleton's unfulfilled ambition to cross Antarctica from the Weddell Sea to the Ross Sea; in the event it was to steal a large share of the limelight.

United States and Soviet activities during the IGY

Complex as these advance plans were for concerting the joint scientific efforts of the co-operating nations in Antarctica, they were dwarfed in magnitude by the logistical problems facing the two countries establishing stations on the ice-sheet. Men and equipment had to be shipped on a scale hitherto undreamed of; construction of the bases had to be carried out in the brief Antarctic spring and summer (October–March) when almost perpetual daylight enables men to work in shifts round the clock. We can better appreciate the scale of these problems by following the activities of the Americans and the Russians in their mammoth efforts to establish their respective stations in time for the beginning of the IGY.

Operations began in the summer season of 1954–55 when the U.S.A. got off to an early start by sending the icebreaker *Atka* on a preliminary reconnaissance to find suitable sites for their bases. They found that Byrd's old base, Little America, had floated out to sea when the edge of the Ross Ice-Shelf, on which it rested, had calved. The following season the United States Navy sent a much larger force to the Ross Sea consisting of 3 icebreakers, 3 cargo vessels, an oil tanker and 2 oil barges. This expedition was code-named Operation Deep Freeze and was the first of a series of annual supply expeditions bearing the name which continues to this day. Operation Deep Freeze 1 1955–6 established two new stations. One was on Ross Island near Captain Scott's old hut dating from the *Discovery* Expedition of 1901–4. This was named Naval Air Facility, McMurdo Sound, changed to McMurdo Station in 1961. The second station was at Kainan Bay, 30 miles to the east of Admiral Byrd's former Little America site. Admiral Byrd himself attended the formal opening ceremony. It was his last farewell to Antarctica; he died in 1957. These coastal stations were designed to act as stepping-stones to two further stations planned on the inland ice. One was the South Pole Station, to be airlifted from McMurdo, the other, named Byrd Station, deep in the heart of Byrd Land, was to be hauled by tractors from Little America. On 20 December 1955 Antarctic aviation history was made when four United States Navy aircraft flew to McMurdo from New Zealand in 14½ hours. This was the first time that heavy cargo planes had linked Antarctica with the rest of the world; today such flights are part of a regular schedule.

The Antarctic season of 1956–7 witnessed another invasion of Antarctica on the grand scale. Operation Deep Freeze 2 numbered 12 ships and over 3000 men. The main task, in the final season before the beginning of the IGY, was to establish the stations at the South Pole and in Byrd Land. The problems involved in transporting the materials for constructing the South Pole Station were formidable; many tons of equipment, including prefabricated huts, tractors and a power plant, all brought to McMurdo by ship the previous season, had to be flown over the high Antarctic plateau and air-dropped. But before this could be done an airfield had to be built at McMurdo to receive the eight United States Air Force C-124 Globemasters, the largest transport aircraft then available. No level areas of ground large enough were to be found; a strip had to be bulldozed clear on the floating ice of McMurdo Sound itself. On 31 October Admiral George Dufek, Commander-in-Chief of Operation Deep Freeze, took off from the floating air-strip in an aircraft fatalistically named 'Que Sera Sera'. Five hours later he and his companions stepped out at 90° south, the first to do so since Robert Falcon Scott and his party in January 1912. The following month work began in earnest on Pole Station, or, as it was to be officially designated, Amundsen-Scott Station, in honour of the Norwegian and British conquerors of the Pole. By the end of February 1957 some 760 tons of material had been air-dropped in the course of over 60 sorties, following Scott's old route over the Ross Ice-Shelf and up the Beardmore Glacier. The scientists, under the leadership of Paul Siple, who first went to Antarctica as a Boy Scout with Byrd in 1928–30, were then left to spend the winter alone on the Antarctic plateau preparing the new station for operational use in the following July. One estimate puts the cost of transporting, sheltering and feeding each man at Amundsen-Scott Station at over a million dollars.

Elsewhere in Antarctica more United States Army engineers supported by the Navy and the Air Force were establishing additional stations. The material for the second inland station, Byrd Station, had to be hauled by tractors from Little America, on the Ross Ice-Shelf, over heavily crevassed terrain and on to the inland plateau. It took five weeks to lay this 650 mile trail. Coastal stations were established by sea. One such was Hallett Station on Cape Adare, a joint United States–New Zealand station close to one of Antarctica's largest penguin rookeries. Others included Ellsworth Station on the Filchner Ice-Shelf in the Weddell Sea and Wilkes Station on the Budd Coast of Wilkes Land in Australian Antarctic Territory.

Meanwhile, the Soviet Union, whose sole interest in Antarctica hitherto had been confined to the whaling industry, was about to renew the tradition of scientific curiosity in these regions associated with Bellingshausen's great voyage of 1819–21. In November 1955 the first Soviet Comprehensive

George Dufek

Antarctic Expedition left the Baltic bound for the Knox Coast of Wilkes Land. It consisted of two ice-breakers, the '*Ob* and *Lena*, and a refrigerator ship, with a total complement of 425 men, of whom 92 were to winter in Antarctica. The cargo, 8345 tons of it, included helicopters, tractors, construction material, provisions and other equipment. The expedition's first task was the construction of an observational station on the Knox Coast which the Russians christened Mirny (Peace). From here they planned to move inland on to the ice-sheet, then virtually unexplored in this sector, and carry out their undertaking to establish a station at the Pole of Inaccessibility (Sovetskaya) and at the Geomagnetic Pole (Vostok). This, as it turned out, was to be a long and arduous operation spread over two seasons, 1956–7 and 1957–8; the Russians, unlike the Americans, could not use aircraft on a scale which would enable them to fly in heavy equipment for air-drops; they had to rely on giant caterpillar tractors for hauling men and supplies, reserving their light aircraft for preliminary reconnaissance flights. A typical sledge train would consist of a tractor hauling three sledges, or cabooses, each serving such specialist functions as radio and seismic cabin, kitchen and living-room, and food and fuel stores. The first inland station to be built was 230 miles from Mirny and was called Pionerskaya; the four scientists left to winter here were the first to do so on the inland ice. In the summer of 1957 an attempt was made to achieve the Geomagnetic Pole but because of numerous setbacks the attempt had to be postponed. Eventually in December 1957 a supply base was established at Komsomolskaya, 530 miles inland, and from here a

Presentation of a plaque, commemorating the achievement of the South Pole by Amundsen and Scott, to the commander of Amundsen-Scott Station by representatives of Britain and Norway, 30 October 1961

Russia returns to the Antarctic to prepare for the IGY. Raising the Soviet flag at Mirny Station, 13 February 1956

M.V. *Theron* unloading supplies at the ice edge, Shackleton Station, for the Commonwealth Trans-Antarctic Expedition

U.S. Neptune P2-V transport aircraft flying over the Beardmore Glacier *en route* to the South Pole during the IGY

successful dash to the Geomagnetic Pole was made. The Pole of Inaccessibility proved an even harder nut to crack. In February 1958 a point was reached 260 miles from Komsomolskaya at latitude 89° 16′E. where a meteorological station, Sovetskaya, was made ready. But not till the following December were the Soviet scientists able to reach the most remote point from all the coasts of Antarctica, latitude 82° 06′S., longitude 54° 58′E., a distance of some 1250 miles from their starting-point on the coast at Mirny. Despite the exceedingly low temperatures experienced – among the lowest in Antarctica – and the lack of oxygen on a plateau over 12,000 ft. above sea-level, the Russians managed to set up a scientific observatory called, simply, Pole of Inaccessibility.

When we come to review the results of the IGY we shall see that the Soviet contribution was one of major importance. Of equal significance not only for the success of the IGY itself, but for the whole future of scientific research in Antarctica, was the wholehearted way in which the Soviet scientists co-operated with their Australian hosts in whose sector of Antarctica they were operating. Australian sovereignty was not (and is not) recognized by the Soviet Union, a situation then fraught with possible friction. As it was the Russians were made cordially welcome by the veteran Antarctic explorer Sir Douglas Mawson, and they, for their part, made the installations at Mirny freely open to inspection by Australian visitors from neighbouring bases. A regular exchange of meteorological data between the two expeditions was organized, and information on all disciplines was soon freely flowing between the Russians and all the nations active in promoting the IGY. Soon Russian scientists were being exchanged with Argentine, Australian, British and American scientists – an absolutely unprecedented measure at a time when elsewhere the 'cold war' still raged. Very possibly the better understanding that exists between the Soviet Union and the Western nations today owes something to the personal friendships, the sharing of knowledge, and the sharing of hardships too, in the neutral territory of Antarctica over a decade ago.

Activities of the other IGY participants

While the two great powers were demonstrating in this spectacular way the large-scale use of modern technology in the exploration of the least accessible parts of Antarctica, the ten other participants were equally active in the coastal regions of the continent and on the Antarctic islands, each making their preparations for the beginning of the IGY. The station map shows how the resources available were used to cover the continent systematically and to the best advantage. Greater Antarctica, before the IGY the least explored part of the continent, had three expeditions active in Dronning Maud Land – Norwegian, Belgian and Japanese. The Australian sector was, as we have seen, divided between the U.S.S.R., the U.S.A. and Australia herself, who had already supported permanent

Sir Vivian Fuchs

Sir Edmund Hillary

stations here since 1947. Terre Adélie, too, had been actively explored by France for some years, while in the adjacent Ross Dependency New Zealand set up her first Antarctic research station on Ross Island, close by the growing complex of huts which formed the U.S.A.'s McMurdo Station. Byrd Land, traditionally an American preserve, supported the two bases described earlier. In Lesser Antarctica the much more accessible Antarctic Peninsula seemed positively overcrowded with some 20 or so bases manned by Argentina, Chile and Great Britain. In addition to her traditional stations in the peninsula, Britain also established two new stations on the Weddell Sea coast. One, Halley Bay, on the Brunt Ice-Shelf, was built by an expedition organized by the Royal Society; the other, Shackleton on the Filchner Ice-Shelf, not far from an Argentine and an American station, was a supply base designed to support the British Commonwealth Trans-Antarctic Expedition.

The British Commonwealth Trans-Antarctic Expedition

'The last great land expedition on earth', as the British Commonwealth Trans-Antarctic Expedition came to be called, was very much in the heroic tradition of Scott and Shackleton's classic journeys. Though not strictly a part of the IGY it attracted enormous public interest throughout the world and had the effect of drawing the attention of the man-in-the-street to the scientific significance of the Antarctic. The plan was to fulfil Shackleton's frustrated attempt to cross the continent from the Weddell Sea to the Ross Sea, but this time using the latest designs in tracked vehicles and light aircraft in addition to traditional dogs and sledges. As with Shackleton's expedition, there were to be two parties. First, a Weddell Sea party led by Dr. (now Sir) Vivian Fuchs would travel overland from Shackleton Base by way of South Ice – a small advance base on the inland plateau – to the South Pole, and from there to McMurdo Sound on the Ross Sea. Second, a Ross Sea party, a New Zealand team led by Sir Edmund Hillary, hero of Everest, was to lay advance depots of fuel and food between Scott Base, on Ross Island, and the polar plateau. This task Hillary accomplished so successfully, using modified Ferguson farm tractors, that he found himself with fuel enough to push on to the American South Pole Station where he and his companions arrived on 4 January 1958, the first men to reach 90°S. overland since Amundsen and Scott. Meanwhile Fuchs's party, which had left Shackleton Base on 24 November 1957, found the going less straightforward. The region where the ice-shelf joined the mainland was a maze of crevasses, and the heavy Snocat and Weasel tractors were obliged to crawl behind a dog team which picked for them a safe path through an area as potentially dangerous as a minefield. Moreover, as part of its contribution to the work of the IGY, the crossing party had undertaken to carry out a seismic and gravity traverse of the continent in an attempt to establish the nature of the underlying terrain. Seismic shots and gravity readings had to be made at frequent intervals and these inevitably delayed progress. At one time it seemed as if Fuchs could not hope to reach the Ross Sea before the sea ice had begun to form and the relief ships had sailed for home. There was the much-publicized occasion when Hillary advised Fuchs over the radio to abandon the attempt and overwinter at the Pole. But Fuchs pressed on to a great ovation at the South Pole Station. After a brief rest the expedition, now without dogs, hastened along its last lap, following the route well marked by Hillary. They made such good time that Scott Base was reached on 2 March 1958 after completing a journey of 2180 miles in 99 days.

Some results of the IGY

What did this tremendous spate of scientific activity in Antarctica achieve? To answer for the many disciplines studied would take a volume in itself. The results for such a relatively short period of time could scarcely be more than provisional, but one or two examples will serve to demonstrate the emergence of new concepts of Antarctica that were to change radically the direction of research in several fields. Take, for example, the old problem of the nature of Antarctica – was it a true continent under the ice, or a group of islands or a combination of both? The traverses, which were the highlight of the IGY, combining as they did both high adventure and novel scientific techniques, were able to cast some light in this direction. The results of the Russian ice soundings in Greater Antarctica all pointed to the bedrock in that region being truly continental in character and above sea-level. These findings were also supported by Australian and French traverses, by American and other Russian soundings at the South Pole, and by the Commonwealth Trans-Antarctic Expedition survey. But in Lesser Antarctica extensive traverses by the Americans suggested that this region was not continental at all and if the ice were to be removed, even allowing for the subsequent rise of the land mass, it would appear as a series of island groups.

Complementing these glaciological studies was the determination of the thickness of the Earth's crust, under the ice-sheet, calculated by studying the times of arrival of seismic disturbances propagated from centres in the Indian Ocean. These data tended to confirm the traverse results in Greater Antarctica; the thickness of the crust here was consistent with that for continents but not for oceans.

Successful studies were also made of the aurora, using all-sky cameras, and a correlation between auroral events in both the Arctic and the Antarctic was discovered. The use of rockets allowed the Antarctic ionosphere to be more thoroughly explored and it was found to exhibit very different properties from those encountered in lower latitudes. New discoveries in the stratosphere helped towards a deeper understanding of the circulation of the atmosphere not only over Antarctica but over the Southern Hemisphere as a whole.

More significant for the future of Antarctica, and indeed for the world, was the enormous success of the whole machinery of co-operation which worked more smoothly than anyone had dared to hope. We have seen how a complete absence of any political restraint allowed a free exchange of scientific information and scientists. A fine example of this was the International Antarctic Weather Central at the United States' Little America base, which, though set up by American scientists, included meteorologists from Argentina, Australia, France, New Zealand, South Africa and the U.S.S.R., all working together in complete harmony.

Science comes to stay in the Antarctic

It was inconceivable that all this potentially valuable activity should suddenly end on 31 December 1958. Indeed the IGY was hardly under way before it became evident that an extension would be needed. Not only were important scientific discoveries being made on a scale undreamed of, but a considerable investment had been made in station facilities and scientific installations which could continue to give valuable service for many years to come. The first positive step in this direction was taken by the International Council of Scientific Unions, in September 1957, when it established SCAR, the Special (now Scientific) Committee on Antarctic Research, and charged it with the task of organizing further scientific research in Antarctica on an international level. This task SCAR has most successfully carried out to the present day. The membership consists of one delegate from each nation active in the Antarctic, as well as representatives of certain member unions of ICSU; each delegate is able to bring with him to meetings advisers in various scientific disciplines as well as in logistics and communications. The work of drawing up programmes of research and of co-ordinating scientific activities is dealt with by a number of Permanent Working Groups established by SCAR. These groups cover the geophysical disciplines studied during the IGY plus a number of additional subjects such as biology, geology and mapping. From time to time the Working Groups convene meetings at which papers are read, discussed and later published. A regular exchange of information on scientific subjects is effected through the national reports to SCAR submitted annually by each member country. These regularly review work accomplished, list activities planned for the forthcoming Antarctic season, and include a bibliography of publications on Antarctic research.

The Antarctic Treaty*

The advent of SCAR, with its open-ended programme of continuous scientific work, marked the end of the old spasmodic exploration of the Antarctic and the beginning of a permanent occupation. But a major problem remained unresolved, the vexed one of territorial claims. We have already seen how, in the days before the IGY, the scientific climate in Antarctica was poisoned by endless arguments over sovereignty. It was essential for the successful continuation of post-IGY activities that there should be no reversion to the old *status quo*. A plan for internationalizing Antarctica under the auspices of the United Nations had been mooted several years previously by the U.S.A. but had come to nothing. In the event it was the U.S.A. which was to take the initiative on a second and more auspicious occasion. In 1958 President Dwight D. Eisenhower invited representatives of the nations concerned to 'confer with us to seek an effective joint means of keeping Antarctica open to all nations to conduct scientific or peaceful activities there'. The nations concurred and a conference met in Washington in October 1959. After several weeks of negotiation the Antarctic Treaty was signed on 1 December 1959 and was finally ratified on 23 June 1961. The signatories included all the nations active in Antarctica during the IGY and these have since been joined by Poland, Czechoslovakia, Denmark and the Netherlands, making a total of 16 powers.

The provisions of the Antarctic Treaty, which applies to the area south of latitude 60°S., embrace six vital principles. These are:

1. Antarctica is to be used for peaceful purposes only. Military personnel may be employed but only on scientific or essential peaceful work.
2. The freedom of scientific investigation and co-operation which characterized the IGY is to continue.

* For the full text of the Treaty see Appendix B.

3. Scientific observations are to be made freely available and scientific personnel are to be exchanged.

4. All political claims are frozen for the duration of the Treaty [30 years].

5. Nuclear explosions and the tipping of radioactive waste in Antarctica are banned.

6. All stations and equipment are open to the inspection of observers appointed by the nations concerned.

Provision was also made for periodic meetings of the Treaty Powers to exchange information and to consult together on matters of common interest appertaining to Antarctica. Four such consultative meetings have been held since the Treaty came into force which have approved a number of measures, among which those relating to the preservation and conservation of Antarctic plant and animal life are probably the most important and are considered to be more comprehensive than those currently operating in any other major region in the world. The Treaty has proved to be such a model of international co-operation and sanity that it has provided useful precedents for a subsequent international treaty relating to outer space. As Dr. Laurence M. Gould, a former chairman of the United States National Committee on Polar Research of the National Academy of Sciences, has said, 'The Antarctic Treaty is indispensable to the world of science which knows no national or other political boundaries; but it is a document unique in history which may take its place alongside the Magna Carta and other great symbols of man's quest of enlightenment and order'.

'I am hopeful that Antarctica in its symbolic robe of white will shine forth as a continent of peace as nations working together there in the cause of science set an example of international co-operation.' This bust dedicated to Admiral Byrd and looking out over McMurdo Station commemorates the past and looks to the future

Appendix A
FURTHER READING

As in so many other fields there has been an explosion in recent years of the published literature relating to the Antarctic regions. Any general review of this field would be outside the scope of the book. Readers interested in more detailed accounts of work in the Antarctic are directed to the bibliographies listed below. The references which follow to periodicals and books are a small selection only of the many recent publications dealing with the Antarctic, more particularly with Antarctica. For additional information readers are invited to write to the Librarian and Information Officer, the Scott Polar Research Institute, Cambridge, England.

BIBLIOGRAPHIES

1. *Recent Polar Literature*
This is one of the best current guides to the literature of the Antarctic and the Arctic, and is based mainly on the periodicals, books and pamphlets received by the Scott Polar Research Institute. References are accompanied by brief indicative abstracts and are classified by subject and region. This bibliography is published as a supplement to the periodical *Polar Record* (see below) and can be purchased separately.

2. *Antarctic Bibliography*
Sponsored by the Office of Antarctic Programs, National Science Foundation, Washington, D.C., and prepared at the Library of Congress, this bibliography forms a continuing series of abstracts and indexes to current Antarctic literature. Volumes of approximately 2000 entries are published every 18 months or so.

PERIODICAL PUBLICATIONS

1. *Polar Record*
The *Polar Record* is the journal of the Scott Polar Research Institute,

Cambridge, England. The journal has appeared regularly since 1931, first twice and then, since 1955, three times a year. It is the only English-language journal of its type and offers a useful and interesting contribution to the literature of the polar regions.

Each issue contains authoritative articles and notes on a wide variety of subjects of polar interest – reviews of scientific work, techniques of living, travelling and working and historical records. There are also accounts of major field activities. Since 1959, it has included the *SCAR Bulletin*, journal of the Scientific Committee on Antarctic Research of the International Council of Scientific Unions.

The *Polar Record* appears in January, May and October.

2. *Antarctic*

This news bulletin, published quarterly by the New Zealand Antarctic Society, P.O. Box 2110, Wellington, New Zealand, is an invaluable guide to the field activities of all the nations actively engaged in the Antarctic. It also contains occasional scientific articles of general interest and reviews a selection of new books relating to the region.

BOOKS

An invaluable general reference work is:

The Antarctic pilot; comprising the coasts of Antarctica and all islands southward of the usual route of vessels. Third edition. London, Hydrographic Department, 1961.

Though basically a guide for mariners, short chapters on all aspects of the Antarctic continent, and many of the islands, are included.

Similar in scope is:

Sailing directions for Antarctica including the off-lying islands south of latitude 60°S. Second edition. Washington, D.C., Navy Hydrographic Office, 1960.

Two books covering Antarctic research are particularly recommended. Both consist of collections of chapters by specialists:

1. *Antarctic research; a review of British scientific achievement in Antarctica. Edited by Sir Raymond Priestley, Raymond J. Adie and G. de Q. Robin. With foreword by H.R.H. The Duke of Edinburgh.* London, Butterworths, 1964.

2. *Antarctica. Edited by Trevor Hatherton.* London, Methuen, 1965.

Still valuable is the special Antarctic edition of *Scientific American* (Vol. 207, No. 3, 1962) the bulk of which is devoted to the state of Antarctic research.

The best general history of the polar regions is:

The white road; a survey of polar exploration. By L. P. Kirwan. London, Hollis & Carter, 1959. A revised edition, *A history of polar exploration*, was published in the Penguin Book series in 1962.

For an informed and readable account of the International Geophysical

Year of 1957–8, the events which led up to it and the scientific achievements made, there is:

Assault on the unknown; the International Geophysical Year. By Walter Sullivan. New York, London, McGraw-Hill, 1959?

MAPS

1. *Antarctica; prepared by the American Geographical Society for the United States Antarctic Research Program.* Washington, D.C., 1965.
   Scale 1:5,000,000. 42 × 56 inches. (Insets: McMurdo Sound, Victoria Land; Antarctica and adjacent seas; subglacial and submarine relief.)

2. *Antarctica.* [*Prepared by Directorate of Overseas Surveys.*] London, 1963. [DOS (Misc.) 135 (Series 3101).]
   Scale 1:15,000,000. 25 × 27 inches.

3. *British Antarctic Territory (north of 75°S.) with South Georgia and South Sandwich Islands.* [*Prepared by Directorate of Overseas Surveys.*] London, 1963. [DOS 813 (Series 3203).]
   Scale 1:3,000,000. 24 × 27 inches.

4. *Antarctica. Atlas plate 65. Compiled and drawn in the cartographic division of the National Geographic Society for the National Geographic Magazine.* Washington, D.C., 1963.
   Scale 1:9,820,800. 25 × 19 inches. (Insets: Subglacial Antarctica; relation of Antarctica to the surrounding continents; Queen Maud Range; McMurdo Sound.)

PUBLICATIONS REFERRED TO IN THE TEXT

BAGSHAWE, Thomas Wyatt.
Two men in the Antarctic; an expedition to Graham Land 1920–1922. Cambridge, University Press, 1939.

CHERRY-GARRARD, A.
The worst journey in the world. London, Chatto & Windus, 1965.

COOK, James.
A voyage towards the South Pole and round the world . . . . Vols. 1, 2. Fourth edition. London, W. Strahan and T. Cadell, 1784.

COOK, Frederick A.
Through the first Antarctic night 1898–1899; a narrative of the voyage of the 'Belgica' among newly discovered lands and over an unknown sea about the South Pole. London, Heinemann, 1900.

FILDES, Robert.
Journal of a voyage kept on board brig 'Cora' of Liverpool bound to New South Shetland, 1820–21. Unpublished log in the Public Record Office, London. (Ships logs 143, supplementary series 11.)

HOOKER, Joseph Dalton.

The botany of the Antarctic voyage of H.M. discovery ships 'Erebus' and 'Terror' in the years 1839–1843. Flora Antarctica Parts 1, 2. London, Reeve Bros., 1847.

MAWSON, Sir Douglas.

The home of the blizzard; being the story of the Australasian Antarctic Expedition, 1911–1914. Vols. 1, 2. London, William Heinemann, 1915.

MURRAY, G.

The Antarctic manual . . ., London, Royal Geographical Society, 1901, pp. 305–35. Extract from the 'journal of a voyage towards the South Pole on board the brig "Tula", under the command of John Biscoe, with the cutter "Lively" in company'.

ROBERTS, Brian Birley.

'Wildlife conservation in the Antarctic', *Oryx*, Vol. 8, 1965–6.

SCOTT, Robert Falcon.

The voyage of the 'Discovery'. Vols. 1 and 2. London, John Murray, 1905.

Scott's last expedition. Arranged by Leonard Huxley. Vols. 1 and 2. London, Smith, Elder, 1913.

WEBSTER, W. H. B.

Narrative of a voyage to the Southern Atlantic Ocean, in the years 1828, 29, 30, performed in H.M. Sloop 'Chanticleer' under the command of the late Captain Henry Foster by order of the Lords Commissioners of the Admiralty. London, Richard Bentley, 1834.

WILSON, Edward Adrian.

Diary of the 'Discovery' Expedition to the Antarctic regions 1901–1904. Edited from the original mss. in the Scott Polar Research Institute, Cambridge by Ann Savours. London, Blandford Press, 1966.

The South Polar Times. Vol. 1, April to August 1902; Vol. 2, April to August 1903. London, Smith, Elder, 1907.

The South Polar Times. Vol. 3, April to October 1911. London, Smith, Elder, 1914.

Appendix B

THE ANTARCTIC TREATY, 1959

An Antarctic Conference held in Washington in 1958 resulted in the signature of the Antarctic Treaty, which entered into force on 30 April 1962.

Since that date there have been five Consultative Meetings held in accordance with Article IX of the Treaty. The first was held at Canberra in 1961, the second at Buenos Aires in 1962, and the third at Brussels in 1964. The fourth was held at Santiago in November 1966 and the fifth at Paris in 1968.

The text of the Antarctic Treaty is reproduced below.

Text of the Antarctic Treaty

The Governments of Argentina, Australia, Belgium, Chile, the French Republic, Japan, New Zealand, Norway, the Union of South Africa, the Union of Soviet Socialist Republics, the United Kingdom of Great Britain and Northern Ireland, and the United States of America,

Recognizing that it is in the interest of all mankind that Antarctica shall continue forever to be used exclusively for peaceful purposes and shall not become the scene or object of international discord;

Acknowledging the substantial contributions to scientific knowledge resulting from international co-operation in scientific investigation in Antarctica;

Convinced that the establishment of a firm foundation for the continuation and development of such co-operation on the basis of freedom of scientific investigation in Antarctica as applied during the International Geophysical Year accords with the interests of science and the progress of all mankind;

Convinced also that a treaty ensuring the use of Antarctica for peaceful purposes only and the continuance of international harmony in Antarctica will further the purposes and principles embodied in the Charter of the United Nations;

Have agreed as follows:

ARTICLE I

1. Antarctica shall be used for peaceful purposes only. There shall be prohibited, *inter alia*, any measures of a military nature, such as the establishment of military bases and fortifications, the carrying out of military manœuvres, as well as the testing of any type of weapons.

2. The present Treaty shall not prevent the use of military personnel or equipment for scientific research or for any other peaceful purpose.

ARTICLE II

Freedom of scientific investigation in Antarctica and co-operation toward that end, as applied during the International Geophysical Year, shall continue, subject to the provision of the present Treaty.

ARTICLE III

1. In order to promote international co-operation in scientific investigation in Antarctica, as provided for in Article II of the present Treaty, the Contracting Parties agree that, to the greatest extent feasible and practicable:

(*a*) information regarding plans for scientific programs in Antarctica shall be exchanged to permit maximum economy and efficiency of operations;

(*b*) scientific personnel shall be exchanged in Antarctica between expeditions and stations;

(*c*) scientific observations and results from Antarctica shall be exchanged and made freely available.

2. In implementing this Article, every encouragement shall be given to the establishment of co-operative working relations with those Specialized Agencies of the United Nations and other international organizations having a scientific or technical interest in Antarctica.

ARTICLE IV

1. Nothing contained in the present Treaty shall be interpreted as:

(*a*) a renunciation by any Contracting Party of previously asserted rights of or claims to territorial sovereignty in Antarctica;

(*b*) a renunciation or diminution by any Contracting Party of any basis of claim to territorial sovereignty in Antarctica which it may have whether as a result of its activities or those of its nationals in Antarctica, or otherwise;

(*c*) prejudicing the position of any Contracting Party as regards its recognition or non-recognition of any other State's right of or claim or basis of claim to territorial sovereignty in Antarctica.

2. No acts or activities taking place while the present Treaty is in force shall constitute a basis for asserting, supporting or denying a claim to territorial sovereignty in Antarctica or create any rights of sovereignty in Antarctica. No new claim, or enlargement of an existing claim, to territorial sovereignty in Antarctica shall be asserted while the present Treaty is in force.

ARTICLE V

1. Any nuclear explosions in Antarctica and the disposal there of radioactive waste material shall be prohibited.

2. In the event of the conclusion of international agreements concerning the use of nuclear energy, including nuclear explosions and the disposal of radioactive waste material, to which all of the Contracting Parties whose representatives are entitled to participate in the meetings provided for under Article IX are parties, the rules established under such agreements shall apply in Antarctica.

ARTICLE VI

The provision of the present Treaty shall apply to the area south of 60° South Latitude, including all ice-shelves, but nothing in the present Treaty shall prejudice or in any way affect the rights, or the exercise of the rights, of any State under international law with regard to the high seas within that area.

ARTICLE VII

1. In order to promote the objectives and ensure the observance of the provisions of the present Treaty, each Contracting Party whose representatives are entitled to participate in the meetings referred to in Article IX of the Treaty shall have the right to designate observers to carry out any inspection provided for by the present Article. Observers shall be nationals of the Contracting Parties which designate them. The names of observers shall be communicated to every other Contracting Party having the right to designate observers, and like notice shall be given of the termination of their appointment.

2. Each observer designated in accordance with the provisions of paragraph 1 of this Article shall have complete freedom of access at any time to any or all areas of Antarctica.

3. All areas of Antarctica, including all stations, installations and equipment within those areas, and all ships and aircraft at points of discharging or embarking cargoes or personnel in Antarctica, shall be open at all times to inspection by any observers designated in accordance with paragraph 1 of this Article.

4. Aerial observation may be carried out at any time over any or all areas of Antarctica by any of the Contracting Parties having the right to designate observers.

5. Each Contracting Party shall, at the time when the present Treaty enters into force for it, inform the other Contracting Parties, and thereafter shall give them notice in advance, of

(*a*) all expeditions to and within Antarctica, on the part of its ships or nationals, and all expeditions to Antarctica organized in or proceeding from its territory;

(*b*) all stations in Antarctica occupied by its nationals; and

(*c*) any military personnel or equipment intended to be introduced by it into Antarctica subject to the conditions prescribed in paragraph 2 of Article I of the present Treaty.

ARTICLE VIII

1. In order to facilitate the exercise of their functions under the present Treaty, and without prejudice to the respective positions of the Contracting Parties relating to jurisdiction over all other persons in Antarctica, observers designated under paragraph 1 of Article VII and scientific personnel exchanged under sub-paragraph 1 (*b*) of Article III of the Treaty, and members of the staffs accompanying any such persons, shall be subject only to the jurisdiction of the Contracting Party of which they are nationals in respect of all acts or omissions occurring while they are in Antarctica for the purpose of exercising their functions.

2. Without prejudice to the provisions of paragraph 1 of this Article, and pending the adoption of measures in pursuance of sub-paragraph 1 (*e*) of Article IX, the Contracting Parties concerned in any case of dispute with regard to the

exercise of jurisdiction in Antarctica shall immediately consult together with a view to reaching a mutually acceptable solution.

ARTICLE IX

1. Representatives of the Contracting Parties named in the preamble to the present Treaty shall meet at the City of Canberra within two months after the date of entry into force of the Treaty, and thereafter at suitable intervals and places, for the purpose of exchanging information, consulting together on matters of common interest pertaining to Antarctica, and formulating and considering, and recommending to their Governments, measures in furtherance of the principles and objectives of the Treaty, including measures regarding:

(*a*) use of Antarctica for peaceful purposes only;
(*b*) facilitation of scientific research in Antarctica;
(*c*) facilitation of international scientific co-operation in Antarctica;
(*d*) facilitation of the exercise of the rights of inspection provided for in Article VII of the Treaty;
(*e*) questions relating to the exercise of jurisdiction in Antarctica;
(*f*) preservation and conservation of living resources in Antarctica.

2. Each Contracting Party which has become a party to the present Treaty by accession under Article XIII shall be entitled to appoint representatives to participate in the meetings referred to in paragraph 1 of the present Article, during such time as that Contracting Party demonstrates its interest in Antarctica by conducting substantial scientific research activity there, such as the establishment of a scientific station or the despatch of a scientific expedition.

3. Reports from the observers referred to in Article VII of the present Treaty shall be transmitted to the representatives of the Contracting Parties participating in the meetings referred to in paragraph 1 of the present Article.

4. The measures referred to in paragraph 1 of this Article shall become effective when approved by all the Contracting Parties whose representatives were entitled to participate in the meetings held to consider those measures.

5. Any or all of the rights established in the present Treaty may be exercised as from the date of entry into force of the Treaty whether or not any measures facilitating the exercise of such rights have been proposed, considered or approved as provided in this Article.

ARTICLE X

Each of the Contracting Parties undertakes to exert appropriate efforts, consistent with the Charter of the United Nations, to the end that no one engages in any activity in Antarctica contrary to the principles or purposes of the present Treaty.

ARTICLE XI

1. If any dispute arises between two or more of the Contracting Parties concerning the interpretation or application of the present Treaty, those Contracting Parties shall consult among themselves with a view to having the dispute resolved by negotiation, inquiry, mediation, conciliation, arbitration, judicial settlement or other peaceful means of their own choice.

2. Any dispute of this character not so resolved shall, with the consent, in each case, of all parties to the dispute, be referred to the International Court of Justice for settlement; but failure to reach agreement on reference to the International Court shall not absolve parties to the dispute from the responsibility of continuing to seek to resolve it by any of the various peaceful means referred to in paragraph 1 of this Article.

ARTICLE XII

1.—(*a*) The present Treaty may be modified or amended at any time by unanimous agreement of the Contracting Parties whose representatives are entitled to participate in the meetings provided for under Article IX. Any such modification or amendment shall enter into force when the depositary Government has received notice from all such Contracting Parties that they have ratified it.

(*b*) Such modification or amendment shall thereafter enter into force as to any other Contracting Party when notice of ratification by it has been received by the depositary Government. Any such Contracting Party from which no notice of ratification is received within a period of two years from the date of entry into force of the modification or amendment in accordance with the provisions of sub-paragraph 1 (*a*) of this Article shall be deemed to have withdrawn from the present Treaty on the date of the expiration of such period.

2.—(*a*) If after the expiration of thirty years from the date of entry into force of the present Treaty, any of the Contracting Parties whose representatives are entitled to participate in the meetings provided for under Article IX so requests by a communication addressed to the depositary Government, a Conference of all the Contracting Parties shall be held as soon as practicable to review the operation of the Treaty.

(*b*) Any modification or amendment to the present Treaty which is approved at such a Conference by a majority of the Contracting Parties there represented, including a majority of those whose representatives are entitled to participate in the meetings provided for under Article IX, shall be communicated by the depositary Government to all the Contracting Parties immediately after the termination of the Conference and shall enter into force in accordance with the provisions of paragraph 1 of the present Article.

(*c*) If any such modification or amendment has not entered into force in accordance with the provisions of sub-paragraph 1 (*a*) of this Article within a period of two years after the date of its communication to all the Contracting Parties, any Contracting Party may at any time after the expiration of that period give notice to the depositary Government of its withdrawal from the present Treaty; and such withdrawal shall take effect two years after the receipt of the notice by the depositary Government.

ARTICLE XIII

1. The present Treaty shall be subject to ratification by the signatory States. It shall be open for accession by any State which is a Member of the United Nations, or by any other State which may be invited to accede to the Treaty with the consent of all the Contracting Parties whose representatives are entitled to participate in the meetings provided for under Article IX of the Treaty.

2. Ratification of or accession to the present Treaty shall be effected by each State in accordance with its constitutional processes.

3. Instruments of ratification and instruments of accession shall be deposited with the Government of the United States of America, hereby designated as the depositary Government.

4. The depositary Government shall inform all signatory and acceding States of the date of each deposit of an instrument of ratification or accession, and the date of entry into force of the Treaty and of any modification or amendment thereto.

5. Upon the deposit of instruments of ratification by all the signatory States, the present Treaty shall enter into force for those States and for States which have deposited instruments of accession*. Thereafter the Treaty shall enter into force for any acceding State upon the deposit of its instrument of accession.

6. The present Treaty shall be registered by the depositary Government pursuant to Article 102 of the Charter of the United Nations.

ARTICLE XIV

The present Treaty, done in the English, French, Russian and Spanish languages, each version being equally authentic, shall be deposited in the archives of the Government of the United States of America, which shall transmit duly certified copies thereof to the Governments of the signatory and acceding States.

ACCESSIONS

	Date of deposit
Czechoslovakia	14 June 1962
Denmark	20 May 1965
Netherlands	30 March 1967
Poland	8 June 1961

* Entered into force: 23 June 1961.

Appendix C

THE ORGANIZATION OF ANTARCTIC RESEARCH

At an international level co-ordination of scientific activity in the Antarctic is the responsibility of the Scientific Committee on Antarctic Research (SCAR), a scientific committee of the International Council of Scientific Unions. A brief account of SCAR's activities will be found in Chapter 14.

At a national level Antarctic research is financed from government funds and is the responsibility of various agencies. In some cases a single agency deals with both the carrying out of scientific programmes and the logistic organization, that is the provision of transport, supplies and general maintenance. An example is the British Antarctic Survey. In other cases science and logistics are separate responsibilities, as in the U.S.A.

Each nation is represented on SCAR by a national committee, a separate body appointed by the appropriate national academy or institution.

The following list of organizations concerned with Antarctic research is by no means a definitive one; it is intended solely as a guide to the principal government and university agencies in addition to a few privately financed bodies.

ARGENTINA

The main centre of Antarctic research is the Instituto Antártico Argentino which co-ordinates all scientific programmes and comprises the Argentine National Committee of SCAR. The institute carries out its own scientific work as well as giving facilities to universities and other institutes. Scientific research in the Antarctic is also carried out by a number of government organizations such as the Naval Meteorological Service and the Naval Hydrographic Service. Logistic support is provided by the Argentine Navy.

AUSTRALIA

The organization responsible for the administration of Australian Antarctic activities is the Antarctic Division of the Department of External Affairs. The Australian National Committee for Antarctic Research, set up by the Australian Academy of Science, represents Australia on SCAR. Scientific work in the field is carried out by a number of government organizations which collectively form the Australian National Antarctic Research Expeditions (ANARE). Co-operating with ANARE are various government and university departments including the Mawson Institute for Antarctic Research, University of Adelaide. In addition to carrying out its own scientific programmes, ANARE is also responsible for providing all logistic support. Sea transport is provided by means of chartered vessels and air support is given by the Royal Australian Air Force.

BELGIUM

Belgian Antarctic activities have in recent years tended to be carried out in co-operation with the Netherlands. The responsibility for the Belgian programme rests with the Comité Politique Scientifique, a committee of Ministers advised by the Conseil National de la Politique Scientifique. The executive body is the Comité de Gestion des Expéditions Antarctiques Belgo–Néerlandaises which organizes the expeditions and liaises with the Netherlands. Sea transport is provided by chartered vessels. The Belgian representative committee on SCAR is the Comité special belge de la recherche scientifique dans l'Antarctique.

CHILE

The planning and integration of scientific work is under the control of the Instituto Antártico Chileno, an autonomous body attached to the Chilean Ministry for Foreign Affairs. A number of government departments and university institutions are also concerned with the planning and execution of the scientific programme. The body representing Chile on SCAR is the Comité Chileno de Investigaciones Antárticas. Logistic support is provided by the Chilean armed forces.

FRANCE

France's Antarctic and sub-Antarctic territories are jointly designated as Territoires des Terres Australes et Antarctiques Françaises (TAAF) which forms an autonomous body within the Ministre Chargé des Territoires d'Outre-Mer. The detail of the Antarctic science programme is the concern of the French Academy of Sciences through its Comité National Français des Recherches Antarctiques (CNFRA). This Committee is also the official representative on SCAR. Logistic support is the responsibility of TAAF. In Terre Adélie this is carried out by a private organization, Expéditions Polaires Françaises (EPF).

GERMAN FEDERAL REPUBLIC

Though not currently active in the Antarctic there is a keen interest in polar research which finds expression in the Deutsche Gesellschaft für Polarforschung (German Society for Polar Research).

ITALY

This is another country that does not participate in Antarctic research at national level. An active centre of polar information is the Istituto Geografico Polare.

JAPAN

Here the executive agency is the Japanese Antarctic Research Expedition (JARE) within the Ministry of Education. JARE, directed by the Minister of Education, includes representatives of numerous government departments concerned with Antarctic research. Among these are the Department of Polar Research, National Science Museum, which is concerned with logistics, and the Maritime Staff Office of the Defence Agency, which provides ice-breaker support. JARE is supported by two planning committees. The first, the National Antarctic Committee of the Science Council of Japan, provides representation on SCAR and recommends a general scientific programme for Antarctic research. The second, the Special Committee for Antarctic Expeditions, National Science Museum, works out the finer details of the National Antarctic Committee's programmes.

NEW ZEALAND

New Zealand Antarctic research is planned by the Ross Dependency Research Committee appointed by the Minister-in-Charge of Scientific and Industrial Research. This Committee is composed of the representatives of various government departments and universities. As the New Zealand National Committee for Antarctic Research it also participates in SCAR. Detailed planning and logistic support is the task of the Antarctic Division of the New Zealand Department of Scientific and Industrial Research which also takes private and university research in the Ross Dependency under its wing. Transport is provided by the Royal New Zealand Air Force and Navy on a reciprocal basis with the U.S.A.

NORWAY

The organization of Norway's Antarctic activities is the responsibility of the Norwegian Polar Institute (Norsk Polarinstitutt). The National Committee for SCAR is appointed by the Academy of Sciences. Chartered vessels are employed for sea transport.

REPUBLIC OF SOUTH AFRICA

There are two main committees concerned with the Republic's scientific activities in the Antarctic and sub-Antarctic. The first, the Interdepartmental Antarctic Committee, considers recommendations for research. The second, the South African Scientific Committee for Antarctic Research, advises the Council for Scientific and Industrial Research on South African participation in SCAR and submits recommendations for scientific work in the Antarctic. The committee also doubles as the South African National Committee for SCAR. Logistic support is provided by the Antarctic Section of the General Division of the Department of Transport. Sea transport is provided by the ice-breaker *RSA*, and air transport by the South African Air Force.

UNITED KINGDOM

The United Kingdom's scientific and logistic effort in the Antarctic is the responsibility of the British Antarctic Survey (BAS) which is under the control of the Natural Environment Research Council (NERC). BAS maintains a number of units where post-graduate work in geology, geophysics, biology and glaciology is carried out. The Survey provides transport in two vessels strengthened for ice, R.R.S. *John Biscoe* and *Shackleton*. The Royal Navy also gives support in the field with H.M.S. *Endurance*. Liaison with SCAR is maintained through the Royal Society's British National Committee for Antarctic Research which maintains close contact with a number of government departments, such as the Meteorological Office, the Directorate of Overseas Surveys, and the Ministry of Defence, as well as non-government bodies like the Medical Research Council and the Scott Polar Research Institute. The Institute, which is a sub-department within the Department of Geography of the University of Cambridge, was founded in 1920 to commemorate and continue the work of Captain Scott's expeditions. Today it teaches at undergraduate level, carries out post-graduate work in a

The Scott Polar Research Institute

number of disciplines and provides a general information service on all matters relating to the polar regions. With its comprehensive library and archives, modern laboratories and cold chambers, it ranks high among the world's polar institutes.

UNITED STATES OF AMERICA

The body responsible for the management of the United States Antarctic Research Program (USARP) is the National Science Foundation. The National Academy of Sciences advises on the general nature of this programme which also has to meet with the approval of the Antarctic Policy Group, a government agency representing the National Science Foundation and the Departments of State and Defense. The National Academy of Sciences works through a Committee on Polar Research which is composed of leading scientists in the disciplines relevant to Antarctic research. It also represents the U.S.A. on SCAR. The National Science Foundation, an independent Federal government agency, works through the Office of Antarctic Programs (OAP) of the Division of Environmental Sciences. The scientific programme which OAP finances and administers is carried out by scientists from government agencies, universities and private corporations. Among OAP's responsibilities are programme planning and co-ordination as well as the organization of information, including records, maps and specimens. Field support for the United States Antarctic Research Program is provided largely by the United States Navy through the United States Naval Support Force, Antarctica. The Support Force is able to call on other military services, such as the Coast Guard, when necessary. Air support is provided by units of the Army, Navy and Air Force.

No account of the United States' Antarctic activities would be complete without a brief mention of the Arctic Institute of North America. This is a private, non-profit-making research and educational organization, under joint United States/Canadian Control, with offices in Montreal, New York and Washington. Its function is to promote studies and research in all scientific disciplines relating to the North American Arctic although its interest now extends to the Antarctic and indeed to all regions where cold climate is an important factor.

UNION OF SOVIET SOCIALIST REPUBLICS

Three government agencies are responsible for Soviet Antarctic activities:

1. The Arctic and Antarctic Research Institute.
2. The Head Office of the Hydrometeorological Service.
3. The Council of Ministers of the U.S.S.R.

The scientific programmes are the responsibility of the Academy of Sciences of the U.S.S.R. and a number of institutes and university departments. The co-ordination of these programmes is in the hands of the Soviet Committee on Antarctic Research of the Academy of Sciences which also represents the Soviet Union on SCAR. Logistic support is the responsibility of the operational section of the Soviet Antarctic expeditions through the Head Office of the Hydrometeorological Service. Sea transport is provided by the Ministry of Sea Transport and aircraft by the Ministry of Civil Aviation.

A few countries do not support national expeditions in the Antarctic but send scientists there under the auspices of other nations. They include the Netherlands (co-operates with Belgium) and Czechoslovakia, Poland and the German Democratic Republic (co-operate with the U.S.S.R.).

Appendix D

STATIONS OPERATING IN THE ANTARCTIC

(Periodically brought up to date in the *Polar Record*.)

Argentina

General Belgrano, latitude 77° 58′S., longitude 38° 48′W.
Alfarez de Navio Sobral, latitude 81° 04′S., longitude 40° 36′W.
Orcadas, latitude 60° 45′S., longitude 44° 43′W.
Teniente Matienzo, latitude 64° 58′S., longitude 60° 02′W.
Almirante Brown, latitude 64° 53′S., longitude 62° 63′W.
Petrel, latitude 63° 28′S., longitude 56° 17′W.
Esperanza, latitude 63° 24′S., longitude 57°W.

Australia

Macquarie Island, latitude 54° 30′S., longitude 158° 57′E.
Mawson, latitude 67° 36′S., longitude 62° 53′E.
Wilkes, latitude 66° 15′S., longitude 110° 32′E.

Chile

Capitán Arturo Prat, latitude 62° 29′S., longitude 59°38′W.
Presidente Pedro Aguirre Cerda, latitude 62° 56′S., longitude 60° 36′W.
General Bernardo O'Higgins, latitude 63° 19′S., longitude 57° 54′W.

France

Ile de la Possession, Iles Crozet, latitude 46° 25′S., longitude 51° 52′E.
Ile Amsterdam, latitude 37° 50′S., longitude 77° 34′E.
Port-aux-Français, latitude 49° 21′S., longitude 70° 12′E.
Dumont d'Urville, latitude 66° 40′S., longitude 140° 01′E.

Japan

Syowa, latitude 69° 00′S., longitude 39° 35′E.

New Zealand

Scott Base, latitude 77° 51′S., longitude 166° 46′E.
Campbell Island, latitude 52° 33′S., longitude 169° 09′E.

South Africa

Marion Island, latitude 46° 53′S., longitude 37° 52′E.
Gough Island, latitude 40° 19′S., longitude 9° 51′W.
Sanae, latitude 70° 19′S., longitude 2° 22′W.

United Kingdom

Stonington Island, latitude 68° 11′S., longitude 67° 00′W.
Argentine Islands, latitude 65° 15′S., longitude 64° 15′W.
Signy Island, latitude 60° 43′S., longitude 45° 36′W.
Adelaide, latitude 67° 46′S., longitude 68° 54′W.
Halley Bay, latitude 75° 31′S., longitude 26° 38′W.
Grytviken, South Georgia, latitude 54° 17′S., longitude 36° 30′W.
Stanley, Falkland Islands, latitude 51° 45′S., longitude 57° 56′W.

U.S.A.

Amundsen-Scott, South Geographical Pole.
New Byrd (or Byrd), latitude 80° 01′S., longitude 119° 32′W.
McMurdo, latitude 77° 51′S., longitude 166° 37′E.
Palmer Station, latitude 64° 46′S., longitude 64° 04′W.
Plateau Station, latitude 79° 28′S., longitude 40° 35′E.

U.S.S.R.

Mirny, latitude 66° 33′S., longitude 93° 01′E.
Novolazarevskaya, latitude 70° 46′S., longitude 11° 50′E.
Molodezhnaya, latitude 67° 40′S., longitude 45° 51′E.
Vostok, latitude 78° 28′S., longitude 106° 48′E.
Bellingshausen, latitude 62° 15′S., longitude 58° 58′W.

INDEX

Bold figures indicate black and white illustrations, and figures prefixed Pl. indicate colour plates.

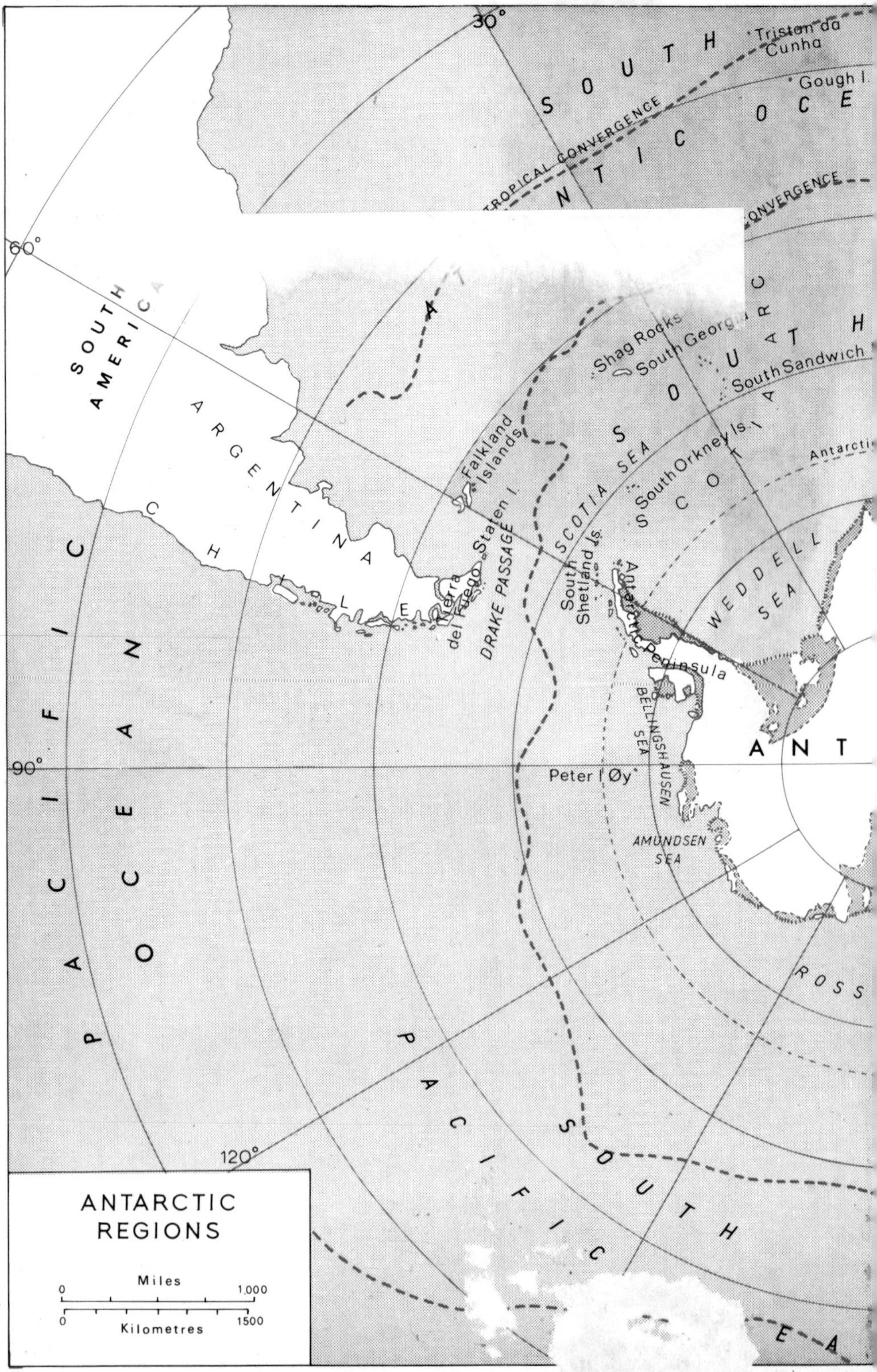

ANTARCTIC REGIONS
Miles
0
1,000
Kilometres
0
1500
30°
60°
90°
120°
SOUTH ATLANTIC OCEAN
Tristan da Cunha
Gough I.
SUB-TROPICAL CONVERGENCE
CONVERGENCE
SOUTH AMERICA
ARGENTINA
CHILE
Falkland Islands
Staten I.
Tierra del Fuego
DRAKE PASSAGE
SCOTIA SEA
Shag Rocks
South Georgia
South Sandwich
South Orkney Is.
South Shetland Is.
Antarctic Peninsula
WEDDELL SEA
BELLINGSHAUSEN SEA
AMUNDSEN SEA
Peter I Øy
PACIFIC OCEAN
SOUTH PACIFIC
SOUTH ATLANTIC
ANT
ROSS